SBAC Math Practice Grade 6

Complete Content Review Plus 2 Full-length SBAC Math Tests

Elise Baniam - Michael Smith

SBAC Math Practice Grade 6

SBAC Math Practice Grade 6
Published in the United State of America By
The Math Notion
Email: info@Mathnotion.com
Web: WWW.MathNotion.com

Copyright © 2020 by the Math Notion. All rights reserved. No part of this publication may be reproduced, stored in a retrieval system, or transmitted in any form or by any means, electronic, mechanical, photocopying, recording, scanning, or otherwise, except as permitted under Section 107 or 108 of the 1976 United States Copyright Ac, without permission of the author.
All inquiries should be addressed to the Math Notion.

ISBN: 978-1-63620-029-3

About the Author

Elise Baniam has been a math instructor for over a decade now. She graduated in Mathematics. Since 2006, Elise has devoted his time to both teaching and developing exceptional math learning materials. As a Math instructor and test prep expert, Elise has worked with thousands of students. She has used the feedback of her students to develop a unique study program that can be used by students to drastically improve their math score fast and effectively.

– SAT Math Workbook
– ACT Math Workbook
– ISEE Math Workbooks
– SSAT Math Workbooks
–many Math Education Workbooks
– and some Mathematics books ...

As an experienced Math teacher, Mrs. Baniam employs a variety of formats to help students achieve their goals: she teaches students in large groups, and she provides training materials and textbooks through her website and through Amazon.

You can contact Elise via email at:
Elise@Mathnotion.com

SBAC Math Practice Grade 6

Get the Targeted Practice You Need to Excel on the Math Section of the SBAC Test Grade 6!

SBAC Math Practice Grade 6 is **an excellent investment in your future** and the best solution for students who want to maximize their score and minimize study time. Practice is an essential part of preparing for a test and improving a test taker's chance of success. The best way to practice taking a test is by going through lots of SBAC math questions.

High-quality mathematics instruction ensures that students become problem solvers. We believe all students can develop deep conceptual understanding and procedural fluency in mathematics. In doing so, through this math workbook we help our students grapple with real problems, think mathematically, and create solutions.

SBAC Math Practice Book allows you to:

- Reinforce your strengths and improve your weaknesses
- Practice **2500+ realistic** SBAC math practice questions
- Exercise math problems in a variety of formats that provide intensive practice
- Review and study **Two Full-length SBAC Practice Tests** with detailed explanations

...and much more!

This Comprehensive SBAC Math Practice Book is carefully designed to provide only that **clear and concise information** you need.

WWW.MathNotion.com

… So Much More Online!

✓ FREE Math Lessons

✓ More Math Learning Books!

✓ Mathematics Worksheets

✓ Online Math Tutors

For a PDF Version of This Book

Please Visit WWW.MathNotion.com

SBAC Math Practice Grade 6

Contents

Chapter 1: Whole Numbers 11
Round Whole Numbers 12
Add and Subtract Integers 13
Multiplication and Division 14
Absolute Value 15
Ordering Integers and Numbers 16
Order of Operations 17
Factoring 18
Great Common Factor (GCF) 19
Least Common Multiple (LCM) 20
Divisibility Rule 21
Answer key Chapter 1 22

Chapter 2: Fractions 27
Adding Fractions – Like Denominator 28
Adding Fractions – Unlike Denominator 29
Subtracting Fractions – Like Denominator 30
Subtracting Fractions – Unlike Denominator 31
Converting Mix Numbers 32
Converting improper Fractions 33
Addition Mix Numbers 34
Subtracting Mix Numbers 35
Simplify Fractions 36
Multiplying Fractions 37
Multiplying Mixed Number 38
Dividing Fractions 39
Dividing Mixed Number 40
Comparing Fractions 41
Answer key Chapter 2 42

Chapter 3: Decimal 47
Round Decimals 48
Decimals Addition 49

SBAC Math Practice Grade 6

    Decimals Subtraction 50

    Decimals Multiplication 51

    Decimal Division... 52

    Comparing Decimals 53

    Convert Fraction to Decimal 54

    Convert Decimal to Percent 55

    Convert Fraction to Percent 56

    Answer key Chapter 3... 57

Chapter 4: Exponent and Radicals... 60

    Positive Exponents 61

    Negative Exponents 62

    Add and subtract Exponents 63

    Exponent multiplication 64

    Exponent division 65

    Scientific Notation 66

    Square Roots 67

    Simplify Square Roots 68

    Answer key Chapter 4... 69

Chapter 5: Ratio, Proportion and Percent.. 72

    Proportions... 73

    Reduce Ratio 74

    Percent 75

    Discount, Tax and Tip 76

    Percent of Change.. 77

    Simple Interest 78

    Answer key Chapter 5... 79

Chapter 6: Measurement 81

    Reference Measurement... 82

    Metric Length Measurement 83

    Customary Length Measurement 83

    Metric Capacity Measurement 84

    Customary Capacity Measurement 84

    Metric Weight and Mass Measurement 85

    Customary Weight and Mass Measurement 85

Unit of Measurements 86

Temperature 87

Time 88

Answers of Worksheets – Chapter 6..89

Chapter 7: Algebraic Expressions 91

Find a Rule...92

Variables and Expressions..93

Translate Phrases 94

Distributive Property 95

Distributive and Simplifying Expressions 96

Factoring Expressions 97

Evaluate One Variable Expressions 98

Evaluate Two Variable Expressions 99

Finding Distance of Two Points...100

Answer key Chapter 7...101

Chapter 8: Equations 105

Graphing Linear Equation 106

One Step Equations..107

Two Steps Equations 108

Multi Steps Equations 109

Answer key Chapter 8..110

Chapter 9: Inequality 112

Graphing Linear Inequalities 113

One Step Inequality..114

Two Steps Inequality 115

Multi Steps Inequality..116

Answer key Chapter 9..117

Chapter 10: Geometry 119

Area and Perimeter of Square 120

Area and Perimeter of Rectangle 121

Area and Perimeter of Triangle..122

Area and Perimeter of Trapezoid 123

Area and Perimeter of Parallelogram...124

Circumference and Area of Circle 125

Perimeter of Polygon 126
Volume of Cubes 127
Volume of Rectangle Prism 128
Volume of Cylinder 129
Surface Area Cubes 130
Surface Area Rectangle Prism 131
Surface Area Cylinder 132
Answer key Chapter 10 133

Chapter 11: Statistics and probability 135

Mean, Median, Mode, and Range 136
Box and Whisker Plot 137
Bar Graph 138
Histogram 139
Dot plots 140
Stem-And-Leaf Plot 141
Pie Graph 142
Probability 143
Answer key Chapter 11 144

SBAC Test Review 149

SBAC Practice Test 1 153
SBAC Practice Test 2 165

Answers and Explanations 177

Answer Key 179
SBAC Practice Test 1 181
SBAC Practice Test 2 187

Chapter 1:
Whole Numbers

Round Whole Numbers

Round to the place of the underlined digit.

1) 8,359,536 ≈ _____

2) 978,245 ≈ _____

3) 5,324,305 ≈ _____

4) 11,246,785 ≈ _____

5) 7,366,552 ≈ _____

6) 4,467,859 ≈ _____

7) 3,458,724 ≈ _____

8) 10,320,668 ≈ _____

9) 12,120,809 ≈ _____

10) 9, 935,890 ≈ _____

11) 10,327,758 ≈ _____

12) 3,147,904 ≈ _____

13) 7,394,885 ≈ _____

14) 5,358,568 ≈ _____

15) 4,386,709 ≈ _____

16) 9,555,665 ≈ _____

17) 2,324,012 ≈ _____

18) 7,425,594 ≈ _____

19) 11,167,780 ≈ _____

20) 9,338,421 ≈ _____

21) 1,846,102 ≈ _____

22) 9,324,489 ≈ _____

Add and Subtract Integers

Find the sum or difference.

1) $(+168) + (+76) =$

2) $(+65) + (-32) =$

3) $217 - 69 =$

4) $(-203) + 179 =$

5) $(-45) + 501 =$

6) $182 + (-265) =$

7) $(-9) + 20 =$

8) $360 - 200 =$

9) $(-10) - (-38) =$

10) $(-67) + (-96) =$

11) $(-143) - 234 =$

12) $1250 - (-346) =$

13) $3 + (-12) + (-20) + (-17) =$

14) $(-28) + (-19) + 31 + 16 =$

15) $(-7) - 11 + 27 - 19 =$

16) $6 + (-20) + (-35 - 24) =$

17) $(+24) + (+32) + (-47) =$

18) $(-35) + (-26) =$

19) $-12 - 17 - 16 - 23 =$

20) $7 + (-21) =$

21) $107 - 80 - 73 - (-38) =$

22) $(20) - (-8) =$

23) $(3) - (5) - (-14) =$

24) $(20) - (6) - (-20) =$

Multiplication and Division

Calculate.

1) $340 \times 8 =$

2) $180 \times 30 =$

3) $(-3) \times 7 \times (-4) =$

4) $-3 \times (-6) \times (-6) =$

5) $12 \times (-12) =$

6) $30 \times (-6) =$

7) $6 \times (-1) \times 5 =$

8) $(-600) \times (-50) =$

9) $(-10) \times (-10) \times 2 =$

10) $165 \times 5 =$

11) $160 \times 80 =$

12) $312 \div 12 =$

13) $(-2,475) \div 3 =$

14) $(-32) \div (-8) =$

15) $384 \div (-24) =$

16) $4,500 \div 36 =$

17) $(-84) \div 2 =$

18) $9,588 \div 6 =$

19) $900 \div (-25) =$

20) $1,680 \div 2 =$

21) $(-81) \div 3 =$

22) $(-1,000) \div (-10) =$

23) $0 \div 250 =$

24) $(-680) \div 4 =$

25) $7,704 \div 856 =$

26) $(-3,150) \div 5 =$

27) $7,268 \div 2 =$

28) $(-48) \div (-4)$

Absolute Value

Simplify each equation below.

1) $|-30| =$

2) $-10 + |-30| + 28 =$

3) $|-48| - |-20| + 12 =$

4) $|-9 + 5 - 3| + |3 + 3| =$

5) $2|2 - 14| + 10 =$

6) $|-6| + |-20| =$

7) $|-36 + 20| + 10 - 9 =$

8) $|-10| - |-23| - 5 =$

9) $|-20| - |-10| + 3 =$

10) $|20| - 28 + |-10| =$

11) $\frac{4|3-6|}{2} =$

12) $|-20 + 9| =$

13) $|-20| \times |5| + 5 =$

14) $|-6| + |-36| + 9 - 3 =$

15) $|-20| + |-20| - 40 =$

16) $13 + |-34 + 15| + |-10| =$

17) $28 - |-63| + 10 =$

18) $\frac{|120|}{|4|} + 6 =$

19) $|-9 + 12| + |32 - 15| + 6 =$

20) $|-20 + 15| + |-5| + 3 =$

21) $\frac{|-32|}{8} \times |-6| =$

22) $\frac{4|4 \times 6|}{2} \times \frac{|-16|}{4} =$

23) $\frac{|2 \times 6|}{12} \times 6 =$

24) $|-10 + 2| \times \frac{|-3 \times 5|}{3} =$

25) $|-100 + 8| - 5 + 5 =$

26) $|-50 + 40| - 10 =$

Ordering Integers and Numbers

Order each set of integers from least to greatest.

1) $7, -8, -5, -2, 3$

2) $-3, -16, 4, 10, 9$

3) $18, -18, -19, 25, -20$

4) $-9, -35, 15, -7, 42$

5) $47, -52, 28, -55, 34$

6) $88, 36, -29, 67, -44$

Order each set of integers from greatest to least.

7) $12, 18, -10, -12, -4$

8) $29, 36, -14, -26, 69$

9) $75, -26, -18, 47, -7$

10) $58, 72, -16, -12, 94$

11) $-7, 99, -15, -48, 64$

12) $-80, -45, -40, 18, 29$

Order of Operations

Evaluate each expression.

1) $5 + (4 \times 3) =$

2) $12 - (3 \times 5) =$

3) $(16 \times 3) + 10 =$

4) $(15 - 5) - (6 \times 3) =$

5) $22 + (16 \div 2) =$

6) $(16 \times 5) \div 5 =$

7) $(84 \div 4) \times (-2) =$

8) $(9 \times 5) + (35 - 12) =$

9) $60 + (2 \times 2) + 8 =$

10) $(30 \times 5) \div (2 + 1) =$

11) $(-8) + (10 \times 4) + 13 =$

12) $(7 \times 6) - (32 \div 4) =$

13) $(9 \times 8 \div 3) - (10 + 11) =$

14) $(12 + 8 - 15) \times 6 - 3 =$

15) $(30 - 12 + 40) \times (95 \div 5) =$

16) $22 + (20 - (32 \div 2)) =$

17) $(6 + 9 - 5 - 8) + (18 \div 2) =$

18) $(85 - 10) + (10 - 15 + 9) =$

19) $(10 \times 2) + (12 \times 5) - 12 =$

20) $12 + 8 - (32 \times 4) + 30 =$

Factoring

Factor, write prime if prime.

1) 12

2) 26

3) 32

4) 48

5) 60

6) 64

7) 35

8) 30

9) 56

10) 75

11) 25

12) 18

13) 49

14) 15

15) 42

16) 124

17) 56

18) 40

19) 75

20) 20

21) 96

22) 27

23) 72

24) 50

25) 24

26) 88

27) 68

28) 124

Great Common Factor (GCF)

Find the GCF of the numbers.

1) 8, 12

2) 48, 32

3) 42, 18

4) 10, 15

5) 18, 24

6) 16, 12

7) 80, 45

8) 100, 75

9) 64, 8

10) 36, 72

11) 93, 62

12) 15, 90

13) 60, 30

14) 36, 28

15) 18, 45

16) 35, 42

17) 12, 20

18) 90, 120, 20

19) 49, 144

20) 16, 28

21) 14, 8, 21

22) 4, 16, 20

23) 14, 49, 7

24) 21, 12

Least Common Multiple (LCM)

Find the LCM of each.

1) 6, 9

2) 30, 24

3) 8, 4, 6

4) 15, 12

5) 30, 5, 40

6) 45, 15

7) 15, 10, 8

8) 3, 4

9) 10, 20, 25

10) 64, 44

11) 24, 36

12) 108, 64

13) 20, 10, 40

14) 12, 20

15) 45, 9, 3

16) 27, 63

17) 42, 12

18) 20, 45

19) 25, 15

20) 14, 32

21) 16, 18

22) 9, 17

23) 32, 18

24) 16, 12

Divisibility Rule

Apply the divisibility rules to find the factors of each number.

1) 12	2, 3, 4, 5, 6, 9, 10		13) 18	2, 3, 4, 5, 6, 9, 10
2) 326	2, 3, 4, 5, 6, 9, 10		14) 405	2, 3, 4, 5, 6, 9, 10
3) 748	2, 3, 4, 5, 6, 9, 10		15) 945	2, 3, 4, 5, 6, 9, 10
4) 81	2, 3, 4, 5, 6, 9, 10		16) 186	2, 3, 4, 5, 6, 9, 10
5) 891	2, 3, 4, 5, 6, 9, 10		17) 640	2, 3, 4, 5, 6, 9, 10
6) 345	2, 3, 4, 5, 6, 9, 10		18) 150	2, 3, 4, 5, 6, 9, 10
7) 75	2, 3, 4, 5, 6, 9, 10		19) 350	2, 3, 4, 5, 6, 9, 10
8) 450	2, 3, 4, 5, 6, 9, 10		20) 4,520	2, 3, 4, 5, 6, 9, 10
9) 1,325	2, 3, 4, 5, 6, 9, 10		21) 990	2, 3, 4, 5, 6, 9, 10
10) 78	2, 3, 4, 5, 6, 9, 10		22) 368	2, 3, 4, 5, 6, 9, 10
11) 772	2, 3, 4, 5, 6, 9, 10		23) 208	2, 3, 4, 5, 6, 9, 10
12) 162	2, 3, 4, 5, 6, 9, 10		24) 500	2, 3, 4, 5, 6, 9, 10

Answer key Chapter 1

Round whole number

1) 8,360,000
2) 978,000
3) 5,324,000
4) 11,247,000
5) 7,367,000
6) 4,468,000
7) 3,458,700
8) 10,320,700
9) 12,120,800
10) 9,900,000
11) 10,328,000
12) 3,148,000
13) 7,390,000
14) 5,359,000
15) 4,386,700
16) 9,556,000
17) 2,324,000
18) 7,425,590
19) 1,168,000
20) 9,340,000
21) 1,850,000
22) 9,324,000

Add and Subtract Integers

1) 244
2) 33
3) 148
4) −24
5) 456
6) −83
7) 11
8) 160
9) 28
10) −163
11) 377
12) 1,596
13) −46
14) 0
15) −10
16) −73
17) 9
18) −61
19) −68
20) −14
21) −8
22) 28
23) 12
24) 34

Multiplication and Division

1) 2,720
2) 5,400
3) 84
4) −108
5) −144
6) −180
7) −30
8) 30,000
9) 200
10) 825
11) 12,800
12) 26
13) −825
14) 4
15) −16
16) 125
17) −42
18) 1,598
19) −36
20) 840
21) −27
22) 100
23) 0
24) −170
25) 9
26) −630
27) 3,634
28) 12

Absolute Value

1) 30
2) 48
3) 40
4) 13
5) 34
6) 26
7) 17
8) −18
9) 13
10) 2
11) 6
12) 11
13) 105
14) 48
15) 0
16) 42
17) −25
18) 36
19) 26
20) 13
21) 24
22) 192
23) 6
24) 40
25) 92
26) 0

Ordering Integers and Numbers

1) $-8, -5, -2, 3, 7$
2) $-16, -3, 4, 9, 10$
3) $-20, -19, -18, 18, 25$
4) $-35, -9, -7, 15, 42$
5) $-55, -52, 28, 34, 47$
6) $-44, -29, 36, 67, 88$
7) $18, 12, -4, -10, -12$
8) $69, 36, 29, -14, -26$
9) $75, 47, -7, -18, -26$
10) $94, 72, 58, -12, -16$
11) $99, 64, -7, -15, -48$
12) $29, 18, -40, -45, -80$

Order of Operations

1) 17
2) −3
3) 58
4) −8
5) 30
6) 16
7) −42
8) 68
9) 72
10) 50
11) 45
12) 34
13) 3
14) 27
15) 1,102
16) 26
17) 11
18) 79
19) 68
20) −78

Factoring

1) 1,2,3,4,6,12
2) 1,2,13,26
3) 1,2,4,8,16,32
4) 1,2,3,4,6,8,12,16,24,48
5) 1,2,3,4,5,6,10,12,15,20,30,60
6) 1,2,4,8,16,32,64
7) 1,5,7,35
8) 1,2,3,5,6,10,15,30
9) 1,2,4,7,8,14,28,56
10) 1,3,5,15,25,75
11) 1,5,25
12) 1,2,3,6,9,18
13) 1,7,49
14) 1,3,5,15
15) 1,2,3,6,7,14,21,42
16) 1,2,4,31,62,124

17) 1,2,4,7,8,14,28,56
18) 1,2,4,5,8,10,20,40
19) 1,3,5,15,25,75
20) 1,2,4,5,10,20
21) 1,2,3,4,6,8,12,24,32,48,96
22) 1,3,9,27
23) 1,2,3,4,6,8,9,12,18,24,36,72
24) 1,2,5,10,25,50
25) 1,2,3,4,6,8,12,24
26) 1,2,4,8,11,22,44,88
27) 1,2,4,17,34,68
28) 1,2,4,31,62,124

Great Common Factor (GCF)

1) 4
2) 16
3) 6
4) 5
5) 6
6) 4
7) 5
8) 25
9) 8
10) 36
11) 31
12) 15
13) 30
14) 4
15) 9
16) 7
17) 4
18) 10
19) 1
20) 4
21) 1
22) 4
23) 7
24) 3

Least Common Multiple (LCM)

1) 18
2) 120
3) 24
4) 60
5) 120
6) 45
7) 120
8) 12
9) 100
10) 704
11) 72
12) 1,728
13) 40
14) 60
15) 45
16) 189
17) 84
18) 180
19) 75
20) 224
21) 144
22) 153
23) 288
24) 48

Divisibility Rule

1) 12   <u>2</u>, <u>3</u>, <u>4</u>, 5, <u>6</u>, 9, 10
2) 326   <u>2</u>, 3, 4, 5, 6, 9, 10
3) 748   <u>2</u>, 3, <u>4</u>, 5, 6, 9, 10
4) 81   2, <u>3</u>, 4, 5, 6, <u>9</u>, 10
5) 891   2, <u>3</u>, 4, 5, 6, <u>9</u>, 10
6) 345   2, <u>3</u>, 4, <u>5</u>, 6, 9, 10
7) 75   2, <u>3</u>, 4, <u>5</u>, 6, 9, 10
8) 450   <u>2</u>, <u>3</u>, 4, <u>5</u>, <u>6</u>, <u>9</u>, <u>10</u>
9) 1,325   2, 3, 4, <u>5</u>, 6, 9, 10
10) 78   <u>2</u>, <u>3</u>, 4, 5, <u>6</u>, 9, 10
11) 772   <u>2</u>, 3, <u>4</u>, 5, 6, 9, 10
12) 162   <u>2</u>, <u>3</u>, 4, 5, <u>6</u>, <u>9</u>, 10

13) 18   <u>2</u>, <u>3</u>, 4, 5, <u>6</u>, <u>9</u>, 10
14) 405   2, <u>3</u>, 4, <u>5</u>, 6, <u>9</u>, 10
15) 945   2, <u>3</u>, 4, <u>5</u>, 6, <u>9</u>, 10
16) 186   <u>2</u>, <u>3</u>, 4, 5, <u>6</u>, 9, 10
17) 640   <u>2</u>, 3, <u>4</u>, <u>5</u>, 6, 9, <u>10</u>
18) 150   <u>2</u>, <u>3</u>, 4, <u>5</u>, 6, 9, <u>10</u>

19) 350   <u>2</u>, 3, 4, <u>5</u>, 6, 9, <u>10</u>
20) 4,520   <u>2</u>, 3, <u>4</u>, <u>5</u>, 6, 9, <u>10</u>
21) 990   <u>2</u>, <u>3</u>, 4, <u>5</u>, <u>6</u>, <u>9</u>, <u>10</u>
22) 368   <u>2</u>, 3, <u>4</u>, 5, 6, 9, 10
23) 208   <u>2</u>, 3, <u>4</u>, 5, 6, 9, 10
24) 500   <u>2</u>, 3, <u>4</u>, 5, 6, 9, <u>10</u>

Chapter 2:
Fractions

Adding Fractions – Like Denominator

Find each sum.

1) $\frac{1}{4} + \frac{2}{4} =$

2) $\frac{2}{5} + \frac{1}{5} =$

3) $\frac{1}{8} + \frac{2}{8} =$

4) $\frac{4}{11} + \frac{1}{11} =$

5) $\frac{4}{21} + \frac{1}{21} =$

6) $\frac{5}{49} + \frac{6}{49} =$

7) $\frac{2}{7} + \frac{11}{7} =$

8) $\frac{1}{15} + \frac{3}{15} =$

9) $\frac{3}{19} + \frac{6}{19} =$

10) $\frac{1}{13} + \frac{1}{13} =$

11) $\frac{1}{5} + \frac{1}{5} =$

12) $\frac{4}{17} + \frac{6}{17} =$

13) $\frac{2}{20} + \frac{17}{20} =$

14) $\frac{4}{25} + \frac{7}{25} =$

15) $\frac{6}{14} + \frac{3}{14} =$

16) $\frac{12}{30} + \frac{5}{30} =$

17) $\frac{1}{9} + \frac{1}{9} =$

18) $\frac{29}{5} + \frac{3}{5} =$

19) $\frac{18}{6} + \frac{5}{6} =$

20) $\frac{25}{37} + \frac{11}{37} =$

Adding Fractions – Unlike Denominator

Add the fractions and simplify the answers.

1) $\frac{1}{3} + \frac{1}{2} =$

2) $\frac{2}{7} + \frac{2}{3} =$

3) $\frac{3}{6} + \frac{1}{5} =$

4) $\frac{5}{13} + \frac{2}{4} =$

5) $\frac{3}{15} + \frac{2}{5} =$

6) $\frac{16}{56} + \frac{3}{16} =$

7) $\frac{3}{7} + \frac{2}{5} =$

8) $\frac{4}{12} + \frac{2}{5} =$

9) $\frac{6}{13} + \frac{3}{7} =$

10) $\frac{3}{8} + \frac{2}{5} =$

11) $\frac{1}{16} + \frac{4}{6} =$

12) $\frac{5}{24} + \frac{2}{3} =$

13) $\frac{3}{36} + \frac{5}{4} =$

14) $\frac{1}{25} + \frac{2}{5} =$

15) $\frac{7}{49} + \frac{3}{7} =$

16) $\frac{7}{12} + \frac{5}{6} =$

17) $\frac{3}{9} + \frac{2}{5} =$

18) $\frac{3}{45} + \frac{1}{5} =$

19) $\frac{3}{18} + \frac{7}{4} =$

20) $\frac{3}{10} + \frac{1}{4} =$

21) $\frac{3}{64} + \frac{1}{8} =$

22) $\frac{6}{14} + \frac{1}{3} =$

23) $\frac{2}{81} + \frac{1}{3} =$

24) $\frac{6}{15} + \frac{1}{3} =$

Subtracting Fractions – Like Denominator

Find the difference.

1) $\dfrac{5}{3} - \dfrac{2}{3} =$

2) $\dfrac{5}{8} - \dfrac{3}{8} =$

3) $\dfrac{11}{14} - \dfrac{8}{14} =$

4) $\dfrac{13}{3} - \dfrac{7}{3} =$

5) $\dfrac{15}{17} - \dfrac{13}{17} =$

6) $\dfrac{18}{33} - \dfrac{10}{33} =$

7) $\dfrac{8}{25} - \dfrac{2}{25} =$

8) $\dfrac{17}{27} - \dfrac{2}{27} =$

9) $\dfrac{7}{10} - \dfrac{3}{10} =$

10) $\dfrac{24}{35} - \dfrac{4}{35} =$

11) $\dfrac{11}{5} - \dfrac{3}{5} =$

12) $\dfrac{28}{38} - \dfrac{18}{38} =$

13) $\dfrac{5}{6} - \dfrac{1}{6} =$

14) $\dfrac{22}{43} - \dfrac{11}{43} =$

15) $\dfrac{4}{7} - \dfrac{3}{7} =$

16) $\dfrac{18}{29} - \dfrac{15}{29} =$

17) $\dfrac{4}{5} - \dfrac{3}{5} =$

18) $\dfrac{42}{53} - \dfrac{38}{53} =$

19) $\dfrac{8}{31} - \dfrac{3}{31} =$

20) $\dfrac{32}{39} - \dfrac{30}{39} =$

21) $\dfrac{9}{26} - \dfrac{5}{26} =$

22) $\dfrac{31}{46} - \dfrac{27}{46} =$

23) $\dfrac{25}{48} - \dfrac{19}{48} =$

24) $\dfrac{39}{65} - \dfrac{27}{65} =$

Subtracting Fractions – Unlike Denominator

Solve each problem.

1) $\frac{1}{2} - \frac{1}{3} =$

2) $\frac{5}{8} - \frac{2}{5} =$

3) $\frac{5}{6} - \frac{2}{7} =$

4) $\frac{3}{5} - \frac{1}{10} =$

5) $\frac{3}{5} - \frac{5}{12} =$

6) $\frac{5}{8} - \frac{5}{16} =$

7) $\frac{2}{25} - \frac{1}{15} =$

8) $\frac{3}{4} - \frac{13}{18} =$

9) $\frac{8}{5} - \frac{7}{6} =$

10) $\frac{5}{6} - \frac{2}{24} =$

11) $\frac{3}{4} - \frac{5}{36} =$

12) $\frac{1}{5} - \frac{2}{25} =$

13) $\frac{7}{6} - \frac{3}{18} =$

14) $\frac{7}{6} - \frac{5}{12} =$

15) $\frac{3}{5} - \frac{2}{9} =$

16) $\frac{3}{5} - \frac{1}{45} =$

17) $\frac{5}{32} - \frac{5}{48} =$

18) $\frac{2}{3} - \frac{2}{7} =$

19) $\frac{3}{5} - \frac{1}{6} =$

20) $\frac{3}{4} - \frac{5}{13} =$

Converting Mix Numbers

Convert the following mixed numbers into improper fractions.

1) $2\frac{3}{4} =$

2) $4\frac{12}{65} =$

3) $9\frac{3}{7} =$

4) $3\frac{5}{6} =$

5) $6\frac{6}{7} =$

6) $2\frac{10}{24} =$

7) $6\frac{7}{12} =$

8) $2\frac{12}{13} =$

9) $2\frac{12}{10} =$

10) $8\frac{6}{7} =$

11) $6\frac{1}{2} =$

12) $5\frac{14}{16} =$

13) $4\frac{8}{7} =$

14) $2\frac{9}{12} =$

15) $8\frac{3}{5} =$

16) $3\frac{4}{12} =$

17) $6\frac{3}{7} =$

18) $2\frac{1}{15} =$

19) $3\frac{7}{15} =$

20) $4\frac{3}{4} =$

21) $3\frac{5}{9} =$

22) $2\frac{11}{5} =$

23) $5\frac{13}{3} =$

24) $11\frac{7}{13} =$

Converting improper Fractions

Convert the following improper fractions into mixed numbers

1) $\frac{67}{12} =$

2) $\frac{75}{63} =$

3) $\frac{19}{15} =$

4) $\frac{58}{45} =$

5) $\frac{85}{26} =$

6) $\frac{271}{52} =$

7) $\frac{84}{63} =$

8) $\frac{41}{5} =$

9) $\frac{16}{15} =$

10) $\frac{11}{2} =$

11) $\frac{35}{4} =$

12) $\frac{120}{95} =$

13) $\frac{120}{54} =$

14) $\frac{28}{8} =$

15) $\frac{83}{11} =$

16) $\frac{31}{3} =$

17) $\frac{101}{8} =$

18) $\frac{51}{48} =$

19) $\frac{28}{9} =$

20) $\frac{8}{7} =$

21) $\frac{7}{2} =$

22) $\frac{43}{10} =$

23) $\frac{32}{24} =$

24) $\frac{78}{7} =$

Addition Mix Numbers

Add the following fractions.

1) $2\frac{1}{3} + 3\frac{1}{3} =$

2) $6\frac{3}{4} + 2\frac{1}{4} =$

3) $1\frac{1}{7} + 2\frac{2}{7} =$

4) $3\frac{1}{6} + 2\frac{3}{2} =$

5) $3\frac{4}{12} + 3\frac{3}{10} =$

6) $4\frac{1}{7} + 1\frac{1}{2} =$

7) $1\frac{2}{21} + 1\frac{2}{24} =$

8) $3\frac{2}{5} + 1\frac{3}{2} =$

9) $2\frac{3}{5} + 2\frac{1}{5} =$

10) $2\frac{4}{5} + 1\frac{3}{5} =$

11) $3\frac{2}{3} + 1\frac{3}{4} =$

12) $4\frac{1}{6} + 1\frac{3}{7} =$

13) $4\frac{1}{2} + 1\frac{3}{2} =$

14) $5\frac{3}{8} + 2\frac{1}{3} =$

15) $2\frac{3}{4} + 3\frac{1}{3} =$

16) $3\frac{1}{4} + 2\frac{3}{5} =$

17) $2\frac{3}{4} + 8\frac{2}{5} =$

18) $1\frac{3}{4} + 1\frac{1}{2} =$

19) $2\frac{3}{4} + 1\frac{1}{7} =$

20) $1\frac{2}{3} + 1\frac{3}{4} =$

21) $3\frac{1}{6} + 2\frac{1}{4} =$

22) $8\frac{2}{5} + 2\frac{3}{4} =$

23) $4\frac{2}{3} + 5\frac{1}{7} =$

24) $2\frac{1}{3} + 3\frac{2}{5} =$

Subtracting Mix Numbers

Subtract the following fractions.

1) $4\frac{1}{2} - 3\frac{1}{2} =$

2) $3\frac{3}{7} - 3\frac{1}{7} =$

3) $6\frac{3}{5} - 5\frac{1}{5} =$

4) $3\frac{1}{3} - 2\frac{1}{2} =$

5) $4\frac{1}{5} - 3\frac{1}{2} =$

6) $9\frac{1}{3} - 5\frac{2}{3} =$

7) $5\frac{5}{10} - 1\frac{6}{10} =$

8) $7\frac{4}{9} - 5\frac{8}{9} =$

9) $6\frac{2}{11} - 5\frac{5}{11} =$

10) $6\frac{2}{5} - 1\frac{1}{5} =$

11) $9\frac{1}{2} - 5\frac{1}{4} =$

12) $2\frac{5}{8} - 1\frac{3}{8} =$

13) $5\frac{3}{58} - 2\frac{5}{6} =$

14) $5\frac{1}{4} - 3\frac{1}{2} =$

15) $17\frac{1}{8} - 12\frac{3}{8} =$

16) $3\frac{3}{5} - 2\frac{1}{5} =$

17) $2\frac{1}{3} - 1\frac{2}{3} =$

18) $2\frac{1}{6} - 1\frac{2}{3} =$

19) $3\frac{2}{6} - 2\frac{1}{2} =$

20) $2\frac{5}{3} - 2\frac{1}{12} =$

21) $2\frac{9}{10} - 1\frac{1}{5} =$

22) $4\frac{2}{5} - 3\frac{1}{11} =$

23) $2\frac{1}{2} - 1\frac{1}{6} =$

24) $2\frac{3}{10} - 1\frac{4}{10} =$

Simplify Fractions

Reduce these fractions to lowest terms

1) $\dfrac{24}{16} =$

2) $\dfrac{18}{27} =$

3) $\dfrac{12}{15} =$

4) $\dfrac{36}{48} =$

5) $\dfrac{9}{27} =$

6) $\dfrac{15}{35} =$

7) $\dfrac{28}{49} =$

8) $\dfrac{80}{100} =$

9) $\dfrac{9}{81} =$

10) $\dfrac{25}{10} =$

11) $\dfrac{24}{32} =$

12) $\dfrac{20}{60} =$

13) $\dfrac{24}{40} =$

14) $\dfrac{3}{12} =$

15) $\dfrac{14}{49} =$

16) $\dfrac{52}{78} =$

17) $\dfrac{96}{36} =$

18) $\dfrac{48}{180} =$

19) $\dfrac{12}{32} =$

20) $\dfrac{88}{77} =$

21) $\dfrac{160}{320} =$

22) $\dfrac{24}{124} =$

23) $\dfrac{144}{36} =$

24) $\dfrac{120}{480} =$

Multiplying Fractions

Find the product.

1) $\frac{2}{7} \times \frac{3}{8} =$

2) $\frac{4}{25} \times \frac{5}{8} =$

3) $\frac{9}{40} \times \frac{10}{27} =$

4) $\frac{6}{13} \times \frac{22}{33} =$

5) $\frac{9}{12} \times \frac{3}{5} =$

6) $\frac{12}{17} \times \frac{5}{3} =$

7) $\frac{5}{6} \times \frac{6}{5} =$

8) $\frac{35}{89} \times 0 =$

9) $\frac{9}{4} \times \frac{12}{5} =$

10) $\frac{10}{18} \times \frac{3}{5} =$

11) $\frac{36}{25} \times \frac{25}{36} =$

12) $\frac{3}{36} \times \frac{6}{27} =$

13) $\frac{15}{7} \times \frac{3}{5} =$

14) $\frac{6}{7} \times \frac{3}{5} =$

15) $\frac{27}{14} \times \frac{7}{3} =$

16) $\frac{12}{17} \times 0 =$

17) $\frac{7}{11} \times \frac{33}{14} =$

18) $\frac{20}{9} \times \frac{3}{5} =$

19) $\frac{9}{16} \times \frac{4}{81} =$

20) $\frac{4}{23} \times \frac{2}{32} =$

21) $\frac{2}{12} \times \frac{3}{16} =$

22) $\frac{25}{8} \times \frac{2}{125} =$

23) $\frac{9}{16} \times \frac{4}{81} =$

24) $\frac{100}{200} \times \frac{400}{800} =$

Multiplying Mixed Number

Multiply. Reduce to lowest terms.

1) $1\frac{2}{3} \times 1\frac{1}{4} =$

2) $1\frac{2}{5} \times 1\frac{3}{2} =$

3) $1\frac{2}{3} \times 3\frac{1}{8} =$

4) $2\frac{1}{8} \times 1\frac{3}{5} =$

5) $2\frac{2}{3} \times 3\frac{1}{3} =$

6) $2\frac{1}{3} \times 1\frac{2}{3} =$

7) $1\frac{3}{4} \times 2\frac{1}{2} =$

8) $3\frac{2}{3} \times 2\frac{1}{3} =$

9) $2\frac{2}{3} \times 2\frac{1}{2} =$

10) $1\frac{1}{3} \times 1\frac{1}{2} =$

11) $2\frac{3}{4} \times 2\frac{2}{3} =$

12) $3\frac{2}{5} \times 2\frac{4}{7} =$

13) $1\frac{3}{4} \times 2\frac{1}{2} =$

14) $1\frac{1}{2} \times 3\frac{1}{7} =$

15) $1\frac{1}{2} \times 2\frac{1}{5} =$

16) $1\frac{2}{7} \times 2\frac{2}{3} =$

17) $1\frac{2}{3} \times 2\frac{1}{5} =$

18) $1\frac{2}{3} \times 3\frac{2}{5} =$

19) $1\frac{3}{4} \times 2\frac{1}{7} =$

20) $1\frac{1}{3} \times 3\frac{2}{5} =$

21) $1\frac{1}{2} \times 2\frac{1}{6} =$

22) $1\frac{1}{9} \times 1\frac{1}{7} =$

Dividing Fractions

Divide these fractions.

1) $0 \div \frac{1}{5} =$

2) $\frac{6}{12} \div 6 =$

3) $\frac{8}{11} \div \frac{3}{4} =$

4) $\frac{14}{64} \div \frac{2}{8} =$

5) $\frac{3}{19} \div \frac{9}{19} =$

6) $\frac{3}{12} \div \frac{15}{36} =$

7) $9 \div \frac{1}{5} =$

8) $\frac{15}{14} \div \frac{3}{7} =$

9) $\frac{6}{15} \div \frac{1}{14} =$

10) $\frac{2}{13} \div \frac{6}{5} =$

11) $\frac{5}{11} \div \frac{3}{10} =$

12) $\frac{15}{28} \div \frac{3}{7} =$

13) $\frac{7}{16} \div \frac{7}{4} =$

14) $\frac{6}{14} \div \frac{30}{7} =$

15) $\frac{8}{23} \div \frac{2}{23} =$

16) $\frac{9}{32} \div \frac{81}{4} =$

17) $\frac{5}{3} \div \frac{10}{27} =$

18) $8 \div \frac{1}{3} =$

19) $\frac{72}{32} \div \frac{3}{9} =$

20) $\frac{2}{30} \div \frac{8}{5} =$

21) $\frac{2}{9} \div \frac{6}{15} =$

22) $\frac{7}{21} \div \frac{3}{4} =$

Dividing Mixed Number

Divide the following mixed numbers. Cancel and simplify when possible.

1) $2\frac{1}{3} \div 2\frac{1}{2} =$

2) $3\frac{1}{8} \div 2\frac{2}{4} =$

3) $3\frac{1}{2} \div 2\frac{3}{5} =$

4) $2\frac{1}{7} \div 2\frac{1}{2} =$

5) $4\frac{1}{5} \div 2\frac{1}{3} =$

6) $2\frac{5}{9} \div 1\frac{2}{5} =$

7) $2\frac{2}{9} \div 1\frac{1}{2} =$

8) $3\frac{1}{7} \div 2\frac{1}{7} =$

9) $2\frac{1}{9} \div 2\frac{1}{2} =$

10) $3\frac{1}{6} \div 2\frac{2}{3} =$

11) $1\frac{2}{3} \div 5\frac{1}{3} =$

12) $3\frac{1}{9} \div 2\frac{2}{3} =$

13) $3\frac{1}{7} \div 1\frac{1}{11} =$

14) $9\frac{4}{7} \div 4\frac{1}{2} =$

15) $3\frac{3}{4} \div 2\frac{1}{2} =$

16) $2\frac{1}{3} \div 3\frac{2}{5} =$

17) $8\frac{3}{4} \div 2\frac{5}{8} =$

18) $3\frac{1}{3} \div 2\frac{3}{5} =$

19) $3\frac{2}{5} \div 2\frac{1}{2} =$

20) $5\frac{3}{8} \div 2\frac{1}{6} =$

21) $6\frac{1}{2} \div 2\frac{1}{4} =$

22) $4\frac{1}{5} \div 2\frac{1}{7} =$

23) $3\frac{1}{5} \div 2\frac{1}{5} =$

24) $2\frac{1}{7} \div 2\frac{1}{5} =$

Comparing Fractions

Compare the fractions, and write >, < or =

1) $\frac{15}{4}$ ___ $\frac{31}{12}$

2) $\frac{34}{5}$ ___ $\frac{1}{4}$

3) $\frac{3}{6}$ ___ $\frac{7}{5}$

4) $\frac{28}{7}$ ___ $\frac{14}{5}$

5) $\frac{1}{6}$ ___ $\frac{3}{5}$

6) $\frac{11}{7}$ ___ $\frac{15}{9}$

7) $\frac{6}{10}$ ___ $\frac{4}{7}$

8) $\frac{21}{12}$ ___ $\frac{23}{6}$

9) $2\frac{1}{10}$ ___ $5\frac{1}{2}$

10) $4\frac{1}{7}$ ___ $2\frac{1}{6}$

11) $2\frac{1}{3}$ ___ $2\frac{1}{4}$

12) $8\frac{6}{7}$ ___ $8\frac{2}{3}$

13) $1\frac{3}{7}$ ___ $2\frac{5}{3}$

14) $\frac{1}{13}$ ___ $\frac{4}{7}$

15) $\frac{41}{65}$ ___ $\frac{17}{43}$

16) $\frac{65}{200}$ ___ $\frac{65}{92}$

17) $12\frac{1}{2}$ ___ $12\frac{1}{7}$

18) $\frac{1}{2}$ ___ $\frac{1}{4}$

19) $\frac{1}{9}$ ___ $\frac{1}{15}$

20) $\frac{8}{14}$ ___ $\frac{6}{10}$

21) $\frac{5}{25}$ ___ $\frac{8}{56}$

22) $\frac{6}{7}$ ___ $\frac{3}{7}$

23) $1\frac{38}{32}$ ___ $2\frac{3}{16}$

24) $4\frac{18}{5}$ ___ $5\frac{4}{3}$

Answer key Chapter 2

Adding Fractions – Like Denominator

1) $\frac{3}{4}$
2) $\frac{3}{5}$
3) $\frac{3}{8}$
4) $\frac{5}{11}$
5) $\frac{5}{21}$
6) $\frac{11}{49}$
7) $\frac{13}{7}$
8) $\frac{4}{15}$
9) $\frac{9}{19}$
10) $\frac{2}{13}$
11) $\frac{2}{5}$
12) $\frac{10}{17}$
13) $\frac{19}{20}$
14) $\frac{11}{25}$
15) $\frac{9}{14}$
16) $\frac{17}{30}$
17) $\frac{2}{9}$
18) $\frac{32}{5}$
19) $\frac{23}{6}$
20) $\frac{36}{37}$

Adding Fractions – Unlike Denominator

1) $\frac{5}{6}$
2) $\frac{20}{21}$
3) $\frac{7}{10}$
4) $\frac{23}{26}$
5) $\frac{3}{5}$
6) $\frac{53}{112}$
7) $\frac{29}{35}$
8) $\frac{11}{15}$
9) $\frac{81}{91}$
10) $\frac{31}{40}$
11) $\frac{35}{48}$
12) $\frac{7}{8}$
13) $\frac{4}{3}$
14) $\frac{11}{25}$
15) $\frac{4}{7}$
16) $\frac{17}{12}$
17) $\frac{11}{15}$
18) $\frac{4}{15}$
19) $\frac{23}{12}$
20) $\frac{11}{20}$
21) $\frac{11}{64}$
22) $\frac{16}{21}$
23) $\frac{29}{81}$
24) $\frac{11}{15}$

Subtracting Fractions – Like Denominator

1) 1
2) $\frac{1}{4}$
3) $\frac{3}{14}$
4) 2
5) $\frac{2}{17}$
6) $\frac{8}{33}$
7) $\frac{6}{25}$
8) $\frac{5}{9}$
9) $\frac{2}{5}$
10) $\frac{4}{7}$
11) $\frac{8}{5}$
12) $\frac{5}{19}$
13) $\frac{2}{3}$
14) $\frac{11}{43}$
15) $\frac{1}{7}$
16) $\frac{3}{29}$
17) $\frac{1}{5}$
18) $\frac{4}{53}$

19) $\frac{5}{31}$ 21) $\frac{2}{13}$ 23) $\frac{1}{8}$

20) $\frac{2}{39}$ 22) $\frac{2}{23}$ 24) $\frac{12}{65}$

Subtracting Fractions – Unlike Denominator

1) $\frac{1}{6}$ 8) $\frac{1}{36}$ 15) $\frac{17}{45}$

2) $\frac{9}{40}$ 9) $\frac{13}{30}$ 16) $\frac{26}{45}$

3) $\frac{23}{42}$ 10) $\frac{3}{4}$ 17) $\frac{5}{96}$

4) $\frac{1}{2}$ 11) $\frac{11}{18}$ 18) $\frac{8}{21}$

5) $\frac{11}{60}$ 12) $\frac{3}{25}$ 19) $\frac{13}{30}$

6) $\frac{5}{16}$ 13) 1   20) $\frac{19}{52}$

7) $\frac{1}{75}$ 14) $\frac{3}{4}$

Converting Mix Numbers

1) $\frac{11}{4}$ 9) $\frac{32}{10}$ 17) $\frac{45}{7}$

2) $\frac{272}{65}$ 10) $\frac{62}{7}$ 18) $\frac{31}{15}$

3) $\frac{66}{7}$ 11) $\frac{13}{2}$ 19) $\frac{52}{15}$

4) $\frac{23}{6}$ 12) $\frac{94}{16}$ 20) $\frac{19}{4}$

5) $\frac{48}{7}$ 13) $\frac{36}{7}$ 21) $\frac{32}{9}$

6) $\frac{58}{24}$ 14) $\frac{33}{12}$ 22) $\frac{21}{5}$

7) $\frac{79}{12}$ 15) $\frac{43}{5}$ 23) $\frac{28}{3}$

8) $\frac{38}{13}$ 16) $\frac{40}{12}$ 24) $\frac{150}{13}$

Converting improper Fractions

1) $5\frac{7}{12}$ 6) $5\frac{11}{52}$ 11) $8\frac{3}{4}$

2) $1\frac{21}{63}$ 7) $1\frac{21}{63}$ 12) $1\frac{25}{95}$

3) $1\frac{4}{15}$ 8) $8\frac{1}{5}$ 13) $2\frac{12}{54}$

4) $1\frac{13}{45}$ 9) $1\frac{1}{15}$ 14) $3\frac{4}{8}$

5) $3\frac{7}{26}$ 10) $5\frac{1}{2}$ 15) $7\frac{6}{11}$

16) $10\frac{1}{3}$ 19) $3\frac{1}{9}$ 22) $4\frac{3}{10}$

17) $12\frac{5}{8}$ 20) $1\frac{1}{7}$ 23) $1\frac{1}{3}$

18) $1\frac{1}{16}$ 21) $3\frac{1}{2}$ 24) $11\frac{1}{7}$

Adding Mix Numbers

1) $5\frac{2}{3}$ 9) $4\frac{4}{5}$ 17) $11\frac{3}{20}$

2) 9 10) $4\frac{2}{5}$ 18) $3\frac{1}{4}$

3) $3\frac{3}{7}$ 11) $5\frac{5}{12}$ 19) $3\frac{25}{28}$

4) $6\frac{2}{3}$ 12) $5\frac{25}{42}$ 20) $3\frac{5}{12}$

5) $6\frac{19}{30}$ 13) 7 21) $5\frac{5}{12}$

6) $5\frac{9}{14}$ 14) $7\frac{17}{24}$ 22) $11\frac{3}{20}$

7) $2\frac{5}{28}$ 15) $6\frac{1}{12}$ 23) $9\frac{17}{21}$

8) $5\frac{9}{10}$ 16) $5\frac{17}{20}$ 24) $5\frac{11}{15}$

Subtracting Mix Numbers

1) 1 9) $\frac{8}{11}$ 17) $\frac{2}{3}$

2) $\frac{2}{7}$ 10) $5\frac{1}{5}$ 18) $\frac{1}{2}$

3) $1\frac{2}{5}$ 11) $4\frac{1}{4}$ 19) $\frac{5}{6}$

4) $\frac{5}{6}$ 12) $1\frac{1}{4}$ 20) $1\frac{7}{12}$

5) $\frac{7}{10}$ 13) $2\frac{19}{87}$ 21) $1\frac{7}{10}$

6) $3\frac{2}{3}$ 14) $1\frac{3}{4}$ 22) $1\frac{17}{55}$

7) $3\frac{9}{10}$ 15) $4\frac{3}{4}$ 23) $1\frac{1}{3}$

8) $1\frac{5}{9}$ 16) $1\frac{2}{5}$ 24) $\frac{9}{10}$

Simplify Fractions

1) $\frac{3}{2}$ 4) $\frac{3}{4}$ 7) $\frac{4}{7}$

2) $\frac{2}{3}$ 5) $\frac{1}{3}$ 8) $\frac{4}{5}$

3) $\frac{4}{5}$ 6) $\frac{3}{7}$ 9) $\frac{1}{9}$

10) $\frac{5}{2}$
11) $\frac{3}{4}$
12) $\frac{1}{3}$
13) $\frac{3}{5}$
14) $\frac{1}{4}$

15) $\frac{2}{7}$
16) $\frac{2}{3}$
17) $\frac{8}{3}$
18) $\frac{4}{15}$
19) $\frac{3}{8}$

20) $\frac{8}{7}$
21) $\frac{1}{2}$
22) $\frac{6}{31}$
23) 4
24) $\frac{1}{4}$

Multiplying Fractions

1) $\frac{3}{28}$
2) $\frac{1}{10}$
3) $\frac{1}{12}$
4) $\frac{4}{13}$
5) $\frac{9}{20}$
6) $\frac{20}{17}$
7) 1
8) 0
9) $\frac{27}{5}$

10) $\frac{1}{3}$
11) 1
12) $\frac{1}{54}$
13) $\frac{9}{7}$
14) $\frac{18}{35}$
15) $\frac{9}{2}$
16) 0
17) $\frac{3}{2}$
18) $\frac{4}{3}$

19) $\frac{1}{36}$
20) $\frac{1}{92}$
21) $\frac{1}{32}$
22) $\frac{1}{20}$
23) $\frac{1}{36}$
24) $\frac{1}{4}$

Multiplying Mixed Number

1) $2\frac{1}{12}$
2) $3\frac{1}{2}$
3) $5\frac{5}{24}$
4) $3\frac{2}{5}$
5) $8\frac{8}{9}$
6) $3\frac{8}{9}$
7) $4\frac{3}{8}$
8) $8\frac{5}{9}$

9) $6\frac{2}{3}$
10) 2
11) $7\frac{1}{3}$
12) $8\frac{26}{35}$
13) $4\frac{3}{8}$
14) $4\frac{5}{7}$
15) $3\frac{3}{10}$
16) $3\frac{3}{7}$

17) $3\frac{2}{3}$
18) $5\frac{2}{3}$
19) $3\frac{3}{4}$
20) $4\frac{8}{15}$
21) $3\frac{1}{4}$
22) $1\frac{17}{63}$

Dividing Fractions

1) 0

2) $\frac{1}{12}$ 9) $\frac{28}{5}$ 16) $\frac{1}{72}$

3) $\frac{32}{33}$ 10) $\frac{5}{39}$ 17) $\frac{9}{2}$

4) $\frac{7}{8}$ 11) $\frac{50}{33}$ 18) 24

5) $\frac{1}{3}$ 12) $\frac{5}{4}$ 19) $\frac{27}{4}$

6) $\frac{3}{5}$ 13) $\frac{1}{4}$ 20) $\frac{1}{24}$

7) 45   14) $\frac{1}{10}$ 21) $\frac{5}{9}$

8) $\frac{5}{2}$ 15) 4   22) $\frac{4}{9}$

Dividing Mixed Number

1) $\frac{14}{15}$ 9) $\frac{38}{45}$ 17) $3\frac{1}{3}$

2) $1\frac{1}{4}$ 10) $1\frac{3}{16}$ 18) $1\frac{11}{39}$

3) $1\frac{9}{26}$ 11) $\frac{5}{16}$ 19) $1\frac{9}{25}$

4) $\frac{6}{7}$ 12) $1\frac{1}{6}$ 20) $2\frac{25}{52}$

5) $1\frac{4}{5}$ 13) $2\frac{37}{42}$ 21) $2\frac{8}{9}$

6) $1\frac{52}{63}$ 14) $2\frac{8}{63}$ 22) $1\frac{24}{25}$

7) $1\frac{13}{27}$ 15) $1\frac{1}{2}$ 23) $1\frac{5}{11}$

8) $1\frac{7}{15}$ 16) $\frac{35}{51}$ 24) $\frac{75}{77}$

Comparing Fractions

1) >   7) >   13) <   19) >

2) >   8) <   14) <   20) <

3) <   9) <   15) >   21) >

4) >   10) >   16) <   22) >

5) <   11) >   17) >   23) =

6) <   12) >   18) >   24) >

Chapter 3:
Decimal

Round Decimals

Round each number to the correct place value

1) 0.6̲4 =

2) 2.0̲4 =

3) 6.6̲23 =

4) 0.3̲77 =

5) 7̲.707 =

6) 0.08̲9 =

7) 6.2̲4 =

8) 76.76̲0 =

9) 1.62̲9 =

10) 10.3̲858 =

11) 1.0̲9 =

12) 4.2̲32 =

13) 3.2̲43 =

14) 6.05̲20 =

15) 63̲.69 =

16) 37̲.32 =

17) 419̲.078 =

18) 512.6̲55 =

19) 12.36̲2 =

20) 65̲.65 =

21) 3.20̲89 =

22) 37.0̲73 =

23) 126.5̲16 =

24) 0.01̲22 =

25) 0.078̲5 =

26) 5.01̲62 =

27) 23.61̲33 =

28) 8.08̲20 =

Decimals Addition

Add the following.

1) 25.52 + 52.25

2) 0.93 + 0.07

3) 18.96 + 12.87

4) 56.106 + 3.198

5) 6.960 + 5.87

6) 4.148 + 3.231

7) 72.72 + 12.87

8) 56.24 + 23.47

9) 43.06 + 11.87

10) 7.961 + 12.87

11) 18.148 + 12.231

12) 65.98 + 8.37

13) 28.05 + 7.37

14) 125.32 + 3.32

Decimals Subtraction

Subtract the following

1) 8.97 − 2.82

2) 84.02 − 67.57

3) 0.65 − 0.2

4) 9.784 − 7.2

5) 0.784 − 0.05

6) 84.62 − 23.81

7) 121.26 − 78.97

8) 24.36 − 8.38

9) 52.59 − 37.6

10) 5.872 − 0.297

11) 61.43 − 18.8

12) 17.732 − 4.314

13) 23.502 − 2.817

14) 135.35 − 23.56

Decimals Multiplication

Solve.

1) 2.1 × 2.6

2) 8.7 × 5.9

3) 7.06 × 2.05

4) 67.08 × 10

5) 13.08 × 1000

6) 32.06 × 7.8

7) 26.12 × 12.01

8) 4.06 × 7.05

9) 18.06 × 0.05

10) 21.09 × 9.07

11) 14.3 × 15.7

12) 5.12 × 0.03

13) 8.05 × 0.21

14) 12.12 × 5.03

Decimal Division

Dividing Decimals.

1) $7 \div 1{,}000 =$

2) $3 \div 10 =$

3) $2.6 \div 1{,}000 =$

4) $0.01 \div 100 =$

5) $7 \div 49 =$

6) $2 \div 82 =$

7) $3 \div 48 =$

8) $8 \div 120 =$

9) $8 \div 100 =$

10) $0.8 \div 0.72 =$

11) $0.7 \div 0.07 =$

12) $0.9 \div 0.36 =$

13) $0.5 \div 0.35 =$

14) $0.6 \div 0.06 =$

15) $2.07 \div 10 =$

16) $7.6 \div 100 =$

17) $7.38 \div 1{,}000 =$

18) $15.6 \div 4.5 =$

19) $45.2 \div 5 =$

20) $0.3 \div 0.03 =$

21) $8.05 \div 2.5 =$

22) $0.05 \div 0.20 =$

23) $0.7 \div 4.4 =$

24) $0.08 \div 50 =$

25) $4.16 \div 0.8 =$

26) $0.08 \div 384 =$

Comparing Decimals

Write the Correct Comparison Symbol (>, < or =)

1) 1.15 ____ 2.15

2) 0.4 ____ 0.385

3) 12.5 ____ 12.500

4) 4.05 ____ 4.50

5) 0.511 ____ 0.51

6) 0.623 ____ 0.723

7) 8.76 ____ 8.678

8) 3.0069 ____ 3.069

9) 23.042 ____ 23.034

10) 6.11 ____ 6.08

11) 2.22 ____ 2.222

12) 0.06 ____ 0.55

13) 1.204 ____ 1.25

14) 4.92 ____ 4.0952

15) 0.44 ____ 0.044

16) 17.04 ____ 17.040

17) 0.090 ____ 0.80

18) 20.217 ____ 22.1

19) 0.021 ____ 0.201

20) 21.5 ____ 11.8

21) 3.5 ____ 10.9

22) 0.071 ____ 0.0701

23) 4.021 ____ 0.4021

24) 2.5 ____ 0.255

25) 5.2 ____ 0.255

26) 2.05 ____ 2.0500

27) 6.05 ____ 0.655

28) 1.0501 ____ 1.0510

Convert Fraction to Decimal

Write each as a decimal.

1) $\frac{40}{100} =$

2) $\frac{38}{100} =$

3) $\frac{4}{25} =$

4) $\frac{6}{24} =$

5) $\frac{9}{81} =$

6) $\frac{49}{100} =$

7) $\frac{2}{25} =$

8) $\frac{17}{25} =$

9) $\frac{47}{200} =$

10) $\frac{13}{50} =$

11) $\frac{18}{36} =$

12) $\frac{3}{8} =$

13) $\frac{6}{20} =$

14) $\frac{9}{125} =$

15) $\frac{27}{50} =$

16) $\frac{20}{50} =$

17) $\frac{45}{10} =$

18) $\frac{6}{30} =$

19) $\frac{67}{1,000} =$

20) $\frac{1}{10} =$

21) $\frac{7}{20} =$

22) $\frac{4}{100} =$

Convert Decimal to Percent

Write each as a percent.

1) 0.165 =

2) 0.15 =

3) 1.4 =

4) 0.015 =

5) 0.005 =

6) 0.625 =

7) 0.185 =

8) 0.34 =

9) 0.03 =

10) 0.1 =

11) 0.175 =

12) 4.95 =

13) 2.105 =

14) 0.2 =

15) 1.05 =

16) 0.0275 =

17) 0.0015 =

18) 0.720 =

19) 2.25 =

20) 0.333 =

21) 6.175 =

22) 0.326 =

23) 1.8 =

24) 0.5 =

25) 1.5 =

26) 12.5 =

27) 3.05 =

28) 0.01 =

Convert Fraction to Percent

Write each as a percent.

1) $\dfrac{1}{5} =$

2) $\dfrac{5}{4} =$

3) $\dfrac{8}{16} =$

4) $\dfrac{19}{22} =$

5) $\dfrac{14}{20} =$

6) $\dfrac{13}{50} =$

7) $\dfrac{7}{9} =$

8) $\dfrac{13}{20} =$

9) $\dfrac{5}{100} =$

10) $\dfrac{8}{20} =$

11) $\dfrac{3}{25} =$

12) $\dfrac{14}{100} =$

13) $\dfrac{48}{50} =$

14) $\dfrac{32}{50} =$

15) $\dfrac{19}{28} =$

16) $\dfrac{3}{33} =$

17) $\dfrac{24}{44} =$

18) $\dfrac{23}{28} =$

19) $\dfrac{24}{84} =$

20) $\dfrac{5}{50} =$

21) $\dfrac{25}{625} =$

22) $\dfrac{480}{240} =$

Answer key Chapter 3

Round Decimals
1) 0.6
2) 2.0
3) 6.6
4) 0.4
5) 8.0
6) 0.09
7) 6.2
8) 76.76
9) 1.63
10) 10.4
11) 1.1
12) 4.2
13) 3.2
14) 6.05
15) 64.0
16) 37.0
17) 420.0
18) 512.7
19) 12.36
20) 66.0
21) 3.21
22) 37.1
23) 126.5
24) 0.01
25) 0.079
26) 5.02
27) 23.61
28) 8.08

Decimals Addition
1) 77.77
2) 1
3) 31.83
4) 59.304
5) 12.83
6) 7.379
7) 85.59
8) 79.71
9) 54.93
10) 20.831
11) 30.379
12) 74.35
13) 35.42
14) 128.64

Decimals Subtraction
1) 6.15
2) 16.45
3) 0.45
4) 2.584
5) 0.734
6) 60.81
7) 42.29
8) 15.98
9) 14.99
10) 5.575
11) 42.63
12) 13.418
13) 20.685
14) 111.79

Decimals Multiplication
1) 5.46
2) 51.33
3) 14.473
4) 670.8
5) 1,3080
6) 250.068
7) 313.7012
8) 28.623
9) 0.903
10) 191.2863
11) 224.51
12) 0.1536
13) 1.6905
14) 60.9636

Decimal Division
1) 0.007
2) 0.3
3) 0.0026

4) 0.0001
5) 0.142…
6) 0.024….
7) 0.0625
8) 0.0666…
9) 0.08
10) 1.111…
11) 10
12) 2.5
13) 1.4285…
14) 10
15) 0.207
16) 0.076
17) 0.00738
18) 3.4666…
19) 9.04
20) 10
21) 3.22
22) 0.25
23) 0.159…
24) 0.0016
25) 5.2
26) 0.0002

Comparing Decimals
1) <
2) >
3) =
4) <
5) >
6) <
7) >
8) <
9) >
10) >
11) <
12) <
13) <
14) >
15) >
16) =
17) <
18) <
19) <
20) >
21) <
22) >
23) >
24) >
25) >
26) =
27) >
28) <

Convert Fraction to Decimal
1) 0.4
2) 0.38
3) 0.16
4) 0.25
5) 0.11
6) 0.49
7) 0.08
8) 0.68
9) 0.235
10) 0.26
11) 0.5
12) 0.375
13) 0.3
14) 0.072
15) 0.54
16) 0.4
17) 4.5
18) 0.2
19) 0.067
20) 0.1
21) 0.35
22) 0.04

Convert Decimal to Percent
1) 16.5%
2) 15%
3) 140%
4) 1.5%
5) 0.5%
6) 62.5%
7) 18.5%
8) 34%
9) 3%

10) 10%
11) 17.5%
12) 495%
13) 210.5%
14) 20%
15) 105%
16) 2.75%

17) 0.15%
18) 72%
19) 225%
20) 33.3%
21) 617.5%
22) 32.6%
23) 180%

24) 50%
25) 150%
26) 1,250%
27) 305%
28) 1%

Convert Fraction to Percent

1) 20%
2) 125%
3) 50%
4) 86.36%
5) 70%
6) 26%
7) 77.8%
8) 65%

9) 5%
10) 40%
11) 12%
12) 14%
13) 96%
14) 64%
15) 67.9%
16) 9.09%

17) 54.5%
18) 82.14%
19) 28.57%
20) 10%
21) 4%
22) 200%

Chapter 4:
Exponent and Radicals

Positive Exponents

Simplify. Your answer should contain only positive exponents.

1) $2^3 =$

2) $5^3 =$

3) $\frac{2x^5y}{xy} =$

4) $(15x3x)^2 =$

5) $(x^3)^2 =$

6) $(\frac{1}{5})^2 =$

7) $0^6 =$

8) $5 \times 5 \times 5 =$

9) $2 \times 2 \times 2 \times 2 \times 2 =$

10) $(3x^2y)^3 =$

11) $10^3 =$

12) $(2x^2y^4)^3 =$

13) $4 \times 10^3 =$

14) $0.5 \times 0.5 \times 0.5 =$

15) $\frac{1}{2} \times \frac{1}{2} \times \frac{1}{2} =$

16) $3^3 =$

17) $(10x^{10}y^3)^2 =$

18) $2^5 =$

19) $x \times x \times x =$

20) $3 \times 3 \times 3 \times 3 \times 3 =$

21) $(3x^2y^3z)^2 =$

22) $7^0 =$

23) $(12x^5y^{-2})^2 =$

24) $(3x^3y^2)^4 =$

Negative Exponents

Simplify. Leave no negative exponents.

1) $3^{-2} =$

2) $7^{-1} =$

3) $(\frac{1}{5})^{-3} =$

4) $10^{-5} =$

5) $1^{-100} =$

6) $4^{-4} =$

7) $(\frac{1}{2})^{-3} =$

8) $-5y^{-3} =$

9) $(\frac{1}{y^{-4}})^{-2} =$

10) $x^{-\frac{3}{2}} =$

11) $\frac{1}{2^{-5}} =$

12) $3^{-4} =$

13) $2^{-3} =$

14) $15^{-1} =$

15) $20^{-2} =$

16) $x^{-4} =$

17) $(x^3)^{-2} =$

18) $x^{-1} \times x^{-1} \times x^{-1} =$

19) $\frac{1}{2} \times \frac{1}{2} =$

20) $10^{-2} =$

21) $10z^{-2} =$

22) $2^{-5} =$

23) $(-\frac{1}{3})^4 =$

24) $6^0 =$

25) $(\frac{1}{x})^{-4} =$

26) $12^{-2} =$

Add and subtract Exponents

Solve each problem.

1) $3^2 + 2^5 =$

2) $x^6 + x^6 =$

3) $3b^2 - 2b^2 =$

4) $3 + 4^3 =$

5) $8 - 4^2 =$

6) $4 + 7^1 =$

7) $2x^3 + 3x^3 =$

8) $10^2 + 3^5 =$

9) $4^5 - 2^4 =$

10) $5^2 - 6^0 =$

11) $1^2 - 3^0 =$

12) $7^1 + 2^3 =$

13) $6^1 - 5^3 =$

14) $3^3 + 3^3 =$

15) $9^2 - 8^2 =$

16) $0^{73} + 0^{54} =$

17) $2^2 - 3^2 =$

18) $7^3 - 7^1 =$

19) $8^2 - 6^2 =$

20) $4^2 + 3^2 =$

21) $2^3 + 4^3 =$

22) $10 + 3^3 =$

23) $6x^5 + 8x^5 =$

24) $8^0 + 4^2 =$

25) $3^2 + 3^2 =$

26) $10^2 + 5^2 =$

27) $(\frac{1}{2})^2 + (\frac{1}{2})^2 =$

28) $9^2 + 3^2 =$

Exponent multiplication

Simplify each of the following

1) $3^6 \times 3^2 =$

2) $9^2 \times 5^0 =$

3) $6^1 \times 7^3 =$

4) $a^{-3} \times a^{-3} =$

5) $y^{-2} \times y^{-2} \times y^{-2} =$

6) $2^4 \times 3^4 \times 2^{-2} \times 3^{-3} =$

7) $5x^2 y^3 \times 8x^3 y^5 =$

8) $(x^2)^3 =$

9) $(x^2 y^3)^4 \times (x^2 y^4)^{-4} =$

10) $6^3 \times 6^2 =$

11) $a^{2b} \times a^0 =$

12) $2^3 \times 2^4 =$

13) $a^m \times a^n =$

14) $a^n \times b^n =$

15) $6^{-2} \times 3^{-2} =$

16) $5^{12} \times 2^{12} =$

17) $(3^5)^4 =$

18) $\left(\frac{1}{5}\right)^3 \times \left(\frac{1}{5}\right)^2 \times \left(\frac{1}{5}\right)^4 =$

19) $\left(\frac{1}{7}\right)^{32} \times 7^{32} =$

20) $(2m)^{\frac{2}{3}} \times (-3m)^{\frac{2}{3}} =$

21) $(x^2 y^3)^{\frac{1}{5}} \times (x^2 y^2)^{\frac{1}{5}} =$

22) $(a^m b^n)^r =$

23) $(3x^2 y^3)^4 =$

24) $(x^{\frac{1}{2}} y^3)^{\frac{-1}{2}} \times (x^2 y^4)^0 =$

25) $6^3 \times 6^4 =$

26) $32^{\frac{1}{4}} \times 32^{\frac{1}{2}} =$

27) $8^4 \times 2^4 =$

28) $(x^3)^0 =$

SBAC Math Practice Grade 6

Exponent division

Simplify. Your answer should contain only positive exponents.

1) $\dfrac{4^3}{4} =$

2) $\dfrac{25x^3}{x} =$

3) $\dfrac{a^m}{a^n} =$

4) $\dfrac{2x^{-5}}{10x^{-3}} =$

5) $\dfrac{81x^8}{9x^3} =$

6) $\dfrac{11x^6}{4x^7} =$

7) $\dfrac{18x^2}{6y^5} =$

8) $\dfrac{35xy^5}{x^5y^2} =$

9) $\dfrac{2x^5}{7x} =$

10) $\dfrac{36x^3y^7}{4x^4} =$

11) $\dfrac{9x^2}{15x^7y^9} =$

12) $\dfrac{yx^4}{5yx^7} =$

13) $\dfrac{14x^2y}{2xy^2} =$

14) $\dfrac{x^{3.25}}{x^{0.25}} =$

15) $\dfrac{5x^3y}{10xy^2} =$

16) $\dfrac{16ab^2r^9}{8a^3b^4} =$

17) $\dfrac{20x^3}{10x^5} =$

18) $\dfrac{16x^3}{4x^6} =$

19) $\dfrac{5^4}{5^2} =$

20) $\dfrac{x}{x^{12}} =$

21) $\dfrac{10^6}{10^2} =$

22) $\dfrac{2xy^4}{8y^2} =$

23) $\dfrac{12x^5y}{144xy^2} =$

24) $\dfrac{42x^6}{7y^8} =$

Scientific Notation

Write each number in scientific notation.

1) $8{,}100{,}000 =$

2) $50 =$

3) $0.0000008 =$

4) $254{,}000 =$

5) $0.000225 =$

6) $6.5 =$

7) $0.00063 =$

8) $19{,}000{,}000 =$

9) $5{,}000{,}000 =$

10) $85{,}000{,}000 =$

11) $0.0000036 =$

12) $0.00012 =$

13) $0.005 =$

14) $6{,}600 =$

15) $1{,}960 =$

16) $170{,}000 =$

17) $0.115 =$

18) $0.05 =$

19) $0.0033 =$

20) $20{,}000 =$

21) $23{,}000 =$

22) $0.00000102 =$

23) $0.0102 =$

24) $1{,}568 =$

25) $32{,}581 =$

26) $12{,}500 =$

27) $12{,}054 =$

28) $60{,}000 =$

Square Roots

Find the square root of each number.

1) $\sqrt{1} =$

2) $\sqrt{4} =$

3) $\sqrt{16} =$

4) $\sqrt{25} =$

5) $\sqrt{49} =$

6) $\sqrt{81} =$

7) $\sqrt{100} =$

8) $\sqrt{144} =$

9) $\sqrt{121} =$

10) $\sqrt{169} =$

11) $\sqrt{9} =$

12) $\sqrt{36} =$

13) $\sqrt{225} =$

14) $\sqrt{196} =$

15) $\sqrt{256} =$

16) $\sqrt{625} =$

17) $\sqrt{289} =$

18) $\sqrt{1,024} =$

19) $\sqrt{484} =$

20) $\sqrt{361} =$

21) $\sqrt{441} =$

22) $\sqrt{841} =$

23) $\sqrt{729} =$

24) $\sqrt{900} =$

25) $\sqrt{400} =$

26) $\sqrt{3,600} =$

27) $\sqrt{4,900} =$

28) $\sqrt{6,400} =$

Simplify Square Roots

Simplify the following.

1) $\sqrt{72} =$

2) $\sqrt{27} =$

3) $\sqrt{28} =$

4) $\sqrt{44} =$

5) $\sqrt{50} =$

6) $\sqrt{40} =$

7) $10\sqrt{125} =$

8) $5\sqrt{600} =$

9) $\sqrt{18} =$

10) $3\sqrt{32} =$

11) $2\sqrt{5} + 8\sqrt{5} =$

12) $\dfrac{1}{1+\sqrt{2}} =$

13) $\sqrt{20} =$

14) $\dfrac{5}{2-\sqrt{3}} =$

15) $\sqrt{3} \times \sqrt{12} =$

16) $\dfrac{\sqrt{400}}{\sqrt{4}} =$

17) $\dfrac{\sqrt{48}}{\sqrt{16 \times 3}} =$

18) $\sqrt{24y^4} =$

19) $7\sqrt{64a} =$

20) $\sqrt{4+32} + \sqrt{16} =$

21) $\sqrt{90} =$

22) $\sqrt{338} =$

23) $\sqrt{60} =$

24) $\sqrt{75} =$

25) $\sqrt{1,875} =$

26) $\sqrt{32} =$

Answer key Chapter 4

Positive Exponents

1) 8
2) 125
3) $2x^4$
4) $2,025x^4$
5) x^6
6) $\frac{1}{25}$
7) 0
8) 5^3
9) 2^5
10) $27x^6y^3$
11) 1,000
12) $8x^6y^{12}$
13) 4,000
14) 0.5^3
15) $(\frac{1}{2})^3$
16) 27
17) $100x^{20}y^6$
18) 32
19) x^3
20) 3^5
21) $9x^4y^6z^2$
22) 1
23) $\frac{144x^{10}}{y^4}$
24) $81x^{12}y^8$

Negative Exponents

1) $\frac{1}{9}$
2) $\frac{1}{7}$
3) 125
4) $\frac{1}{100,000}$
5) 1
6) $\frac{1}{256}$
7) 8
8) $\frac{-5}{y^3}$
9) y^8
10) $\frac{1}{x^{\frac{3}{2}}}$
11) 2^5
12) $\frac{1}{81}$
13) $\frac{1}{8}$
14) $\frac{1}{15}$
15) $\frac{1}{400}$
16) $\frac{1}{x^4}$
17) $\frac{1}{x^6}$
18) $\frac{1}{x^3}$
19) $\frac{1}{2^2}$
20) $\frac{1}{100}$
21) $\frac{10}{z^2}$
22) $\frac{1}{32}$
23) $\frac{1}{81}$
24) 1
25) x^4
26) $\frac{1}{144}$

Add and subtract Exponents

1) 41
2) $2x^6$
3) b^2
4) 67
5) −8
6) 11
7) $5x^3$
8) 343
9) 1,008
10) 24
11) 0
12) 15
13) −119
14) 54
15) 17
16) 0
17) −5
18) 336

19) 28
20) 25
21) 72
22) 37
23) $14x^5$
24) 17
25) 18
26) 125
27) $\frac{1}{2}$
28) 90

Exponent multiplication

1) 3^8
2) 81
3) 2,058
4) a^{-6}
5) y^{-6}
6) $2^2 \times 3^1 = 12$
7) $40x^5y^8$
8) x^6
9) y^{-4}
10) 6^5
11) a^{2b}
12) 2^7
13) a^{m+n}
14) $(ab)^n$
15) 18^{-2}
16) 10^{12}
17) 3^{20}
18) $(\frac{1}{5})^9$
19) 1
20) $(-6m)^{\frac{2}{3}}$
21) $x^{\frac{4}{5}}y$
22) $a^{mr}b^{nr}$
23) $81x^8y^{12}$
24) $x^{\frac{-1}{4}}y^{\frac{-3}{2}}$
25) 6^7
26) $32^{\frac{3}{4}}$
27) $16^4 = 2^{16}$
28) 1

Exponent division

1) 4^2
2) $25x^2$
3) a^{m-n}
4) $\frac{1}{5x^2}$
5) $9x^5$
6) $\frac{11}{4x}$
7) $\frac{3x^2}{y^5}$
8) $\frac{35y^3}{x^4}$
9) $\frac{2x^4}{7}$
10) $\frac{9y^7}{x}$
11) $\frac{3}{5x^5y^9}$
12) $\frac{1}{5x^3}$
13) $\frac{7x}{y}$
14) x^3
15) $\frac{x^2}{2y}$
16) $\frac{2r^9}{a^2b^2}$
17) $\frac{2}{x^2}$
18) $\frac{4}{x^3}$
19) 5^2
20) $\frac{1}{x^{11}}$
21) 10^4
22) $\frac{1}{4}xy^2$
23) $\frac{x^4}{12y}$
24) $\frac{6x^6}{y^8}$

Scientific Notation

1) 81×10^5
2) 5×10^1
3) 8×10^{-7}
4) 2.54×10^5
5) 2.25×10^{-4}
6) 65×10^{-1}
7) 63×10^{-5}
8) 1.9×10^7
9) 5×10^6

10) 8.5×10^7
11) 3.6×10^{-6}
12) 1.2×10^{-4}
13) 5×10^{-3}
14) 6.6×10^3
15) 1.96×10^3
16) 1.7×10^5
17) 1.15×10^{-1}
18) 5×10^{-2}
19) 33×10^{-4}
20) 2×10^4
21) 23×10^3
22) 102×10^{-8}
23) 1.02×10^{-2}
24) 1.568×10^3
25) 32.581×10^3
26) 12.5×10^3
27) 1.2054×10^4
28) 6×10^4

Square Roots

1) 1
2) 2
3) 4
4) 5
5) 7
6) 9
7) 10
8) 12
9) 11
10) 13
11) 3
12) 6
13) 15
14) 14
15) 16
16) 25
17) 17
18) 32
19) 22
20) 19
21) 21
22) 29
23) 27
24) 30
25) 20
26) 60
27) 70
28) 80

Simplify Square Roots

1) $6\sqrt{2}$
2) $3\sqrt{3}$
3) $2\sqrt{7}$
4) $2\sqrt{11}$
5) $5\sqrt{2}$
6) $2\sqrt{10}$
7) $50\sqrt{5}$
8) $50\sqrt{6}$
9) $3\sqrt{2}$
10) $12\sqrt{2}$
11) $10\sqrt{5}$
12) $\sqrt{2} - 1$
13) $2\sqrt{5}$
14) $10 + 5\sqrt{3}$
15) 6
16) 10
17) 1
18) $2y^2\sqrt{6}$
19) $56\sqrt{a}$
20) 10
21) $3\sqrt{10}$
22) $13\sqrt{2}$
23) $2\sqrt{15}$
24) $5\sqrt{3}$
25) $25\sqrt{3}$
26) $4\sqrt{2}$

Chapter 5: Ratio, Proportion and Percent

Proportions

Find a missing number in a proportion.

1) $\dfrac{5}{8} = \dfrac{20}{a}$

2) $\dfrac{a}{6} = \dfrac{24}{36}$

3) $\dfrac{14}{42} = \dfrac{a}{3}$

4) $\dfrac{15}{a} = \dfrac{75}{32}$

5) $\dfrac{8}{a} = \dfrac{32}{150}$

6) $\dfrac{\sqrt{16}}{5} = \dfrac{a}{30}$

7) $\dfrac{5}{12} = \dfrac{15}{a}$

8) $\dfrac{6}{12} = \dfrac{a}{33.6}$

9) $\dfrac{8}{a} = \dfrac{3.2}{4}$

10) $\dfrac{1}{16} = \dfrac{3}{a}$

11) $\dfrac{10}{8} = \dfrac{5}{a}$

12) $\dfrac{12}{a} = \dfrac{3}{17}$

13) $\dfrac{2}{7} = \dfrac{a}{10}$

14) $\dfrac{\sqrt{25}}{4} = \dfrac{30}{a}$

15) $\dfrac{12}{a} = \dfrac{13.2}{19.8}$

16) $\dfrac{50}{190} = \dfrac{a}{380}$

17) $\dfrac{32}{100} = \dfrac{a}{52}$

18) $\dfrac{27}{81} = \dfrac{a}{3}$

19) $\dfrac{5}{8} = \dfrac{1}{a}$

20) $\dfrac{5}{3} = \dfrac{35}{a}$

Reduce Ratio

Reduce each ratio to the simplest form.

1) 3: 12 =

2) 4: 24 =

3) 81: 45 =

4) 30: 25 =

5) 24: 240 =

6) 80: 10 =

7) 80: 400 =

8) 5: 180 =

9) 24: 72 =

10) 3.6: 4.2 =

11) 220: 660 =

12) 1.8: 3 =

13) 150: 250 =

14) 40: 60 =

15) 26: 52 =

16) 16: 4 =

17) 100: 25 =

18) 10: 100 =

19) 108: 72 =

20) 130: 165 =

21) 30: 60 =

22) 24: 28 =

23) 10: 150 =

24) 15: 90 =

Percent

Find the Percent of Numbers.

1) 20% of 38 =

2) 42% of 7 =

3) 11% of 11 =

4) 36% of 75 =

5) 5% of 50 =

6) 32% of 14 =

7) 12% of 3 =

8) 9% of 47 =

9) 50% of 52 =

10) 7.5% of 60 =

11) 92% of 12 =

12) 80% of 60 =

13) 12% of 120 =

14) 1% of 310 =

15) 32% of 0 =

16) 62% of 100 =

17) 32% of 44 =

18) 15% of 60 =

19) 5% of 10 =

20) 3% of 7 =

21) 40% of 20 =

22) 70% of 2 =

23) 25% of 20 =

24) 7% of 200 =

25) 50% of 300 =

26) 3% of 6 =

27) 6% of 400 =

28) 9% of 6 =

Discount, Tax and Tip

Find the selling price of each item.

1) Original price of a computer: $250
   Tax: 6%, Selling price: $_____

2) Original price of a laptop: $320
   Tax: 5%, Selling price: $_____

3) Original price of a sofa: $400
   Tax: 7%, Selling price: $_____

4) Original price of a car: $16,500
   Tax: 4.5%, Selling price: $_____

5) Original price of a Table: $300
   Tax: 6%, Selling price: $_____

6) Original price of a house: $450,000
   Tax: 2.5%, Selling price: $_____

7) Original price of a tablet: $200
   Discount: 20%, Selling price: $_____

8) Original price of a chair: $250
   Discount: 15%, Selling price: $_____

9) Original price of a book: $50
   Discount: 35% Selling price: $_____

10) Original price of a cellphone: 600
    Discount: 10% Selling price: $_____

11) Food bill: $32
    Tip: 20%   Price: $_____

12) Food bill: $30
    Tipp: 15%   Price: $_____

13) Food bill: $64
    Tip: 20%   Price: $_____

14) Food bill: $36
    Tipp: 25%   Price: $_____

Find the answer for each word problem.

15) Nicolas hired a moving company. The company charged $200 for its services, and Nicolas gives the movers a 30% tip. How much does Nicolas tip the movers? $_____

16) Mason has lunch at a restaurant and the cost of his meal is $60. Mason wants to leave a 10% tip. What is Mason's total bill including tip? $_____

Percent of Change

Find each percent of change.

1) From 200 to 400. ___%

2) From 25 ft to 125 ft. ___%

3) From $50 to $350. ___%

4) From 40 cm to 160 cm. ___%

5) From 20 to 60. ___%

6) From 40 to 8. ___%

7) From 160 to 240. ___%

8) From 600 to 300. ___%

9) From 75 to 45. ___%

10) From 128 to 32. ___%

Calculate each percent of change word problem.

11) Bob got a raise, and his hourly wage increased from $24 to $30. What is the percent increase? ____%

12) The price of a pair of shoes increases from $60 to $96. What is the percent increase? ___%

13) At a coffeeshop, the price of a cup of coffee increased from $2.40 to $2.88. What is the percent increase in the cost of the coffee? ____%

14) 24cm are cut from a 96 cm board. What is the percent decrease in length? _%

15) In a class, the number of students has been increased from 108 to 162. What is the percent increase? ____%

16) The price of gasoline rose from $16.80 to $19.32 in one month. By what percent did the gas price rise? ____%

17) A shirt was originally priced at $24. It went on sale for $19.20. What was the percent that the shirt was discounted? ____%

Simple Interest

Determine the simple interest for these loans.

1) $225 at 14% for 2 years. $ _____
2) $2,600 at 8% for 3 years. $ _____
3) $1,300 at 15% for 5 years. $ _____
4) $8,400 at 2.5% for 5 months. $ ___
5) $300 at 2% for 9 months. $ _____

6) $48,000 at 5.5% for 5 years. $ ____
7) $5,200 at 9% for 2 years. $ _____
8) $600 at 5.5% for 4 years. $ _____
9) $800 at 4.5 % for 9 months. $ ____
10) $6,000 at 2.2% for 5 years. $ ___

Calculate each simple interest word problem.

11) A new car, valued at $14,000, depreciates at 4.5% per year. What is the value of the car one year after purchase? $_____

12) Sara puts $8,000 into an investment yielding 5% annual simple interest; she left the money in for two years. How much interest does Sara get at the end of those two years? $_____

13) A bank is offering 10.5% simple interest on a savings account. If you deposit $22,500, how much interest will you earn in two years? $_____

14) $800 interest is earned on a principal of $8,000 at a simple interest rate of 5% interest per year. For how many years was the principal invested? _____

15) In how many years will $1,500 yield an interest of $300 at 5% simple interest? _____

16) Jim invested $6,000 in a bond at a yearly rate of 3.5%. He earned $630 in interest. How long was the money invested? _____

Answer key Chapter 5

Proportions

1) $a = 32$
2) $a = 4$
3) $a = 1$
4) $a = 6.4$
5) $a = 37.5$
6) $a = 24$
7) $a = 36$
8) $a = 16.8$
9) $a = 10$
10) $a = 48$
11) $a = 4$
12) $a = 68$
13) $a = \frac{20}{7}$
14) $a = 24$
15) $a = 18$
16) $a = 100$
17) $a = 16.64$
18) $a = 1$
19) $a = 1.6$
20) $a = 21$

Reduce Ratio

1) 1: 4
2) 1: 6
3) 9: 5
4) 6: 5
5) 1: 10
6) 8: 1
7) 1: 5
8) 1: 36
9) 1: 3
10) 0.6: 0.7
11) 11: 33
12) 0.6: 1
13) 3: 5
14) 2: 3
15) 1: 2
16) 4: 1
17) 4: 1
18) 1: 10
19) 3: 2
20) 26: 33
21) 1: 2
22) 6: 7
23) 1: 15
24) 1: 6

Percent

1) 7.6
2) 2.94
3) 1.21
4) 27
5) 2.5
6) 4.48
7) 0.36
8) 4.23
9) 26
10) 4.5
11) 11.04
12) 48
13) 14.4
14) 3.1
15) 0
16) 62
17) 14.08
18) 9
19) 0.5
20) 0.21
21) 8
22) 1.4
23) 5
24) 14
25) 150
26) 0.18
27) 24
28) 0.54

Discount, Tax and Tip

1) $265.00
2) $336.00
3) $428.00
4) $17,242.50
5) $318.00
6) $461,250
7) $240.00
8) $287.50
9) $67.50
10) $660.00
11) $38.40
12) $34.50
13) $76.80
14) $45.00
15) $60.00
16) $66.00

Percent of Change

1) 100%
2) 400%
3) 600%
4) 300%
5) 200%
6) 80%
7) 50%
8) 50%
9) 40%
10) 75%
11) 25%
12) 60%
13) 20%
14) 25%
15) 50%
16) 15%
17) 20%

Simple Interest

1) $63.00
2) $624.00
3) $975.00
4) $87.50
5) $4.50
6) $13,200.00
7) $936.00
8) $132.00
9) $27.00
10) $660.00
11) $13,370.00
12) $800.00
13) $4725.00
14) 2 *years*
15) 4 *years*
16) 3 *years*

Chapter 6:

Measurement

Reference Measurement

LENGTH	
Customary	**Metric**
1 mile (mi) = 1,760 yards (yd)	1 kilometer (km) = 1,000 meters (m)
1 yard (yd) = 3 feet (ft)	1 meter (m) = 100 centimeters (cm)
1 foot (ft) = 12 inches (in.)	1 centimeter(cm) = 10 millimeters(mm)
VOLUME AND CAPACITY	
Customary	**Metric**
1 gallon (gal) = 4 quarts (qt)	1 liter (L) = 1,000 milliliters (mL)
1 quart (qt) = 2 pints (pt.)	
1 pint (pt.) = 2 cups (c)	
1 cup (c) = 8 fluid ounces (Fl oz)	
WEIGHT AND MASS	
Customary	**Metric**
1 ton (T) = 2,000 pounds (lb.)	1 kilogram (kg) = 1,000 grams (g)
1 pound (lb.) = 16 ounces (oz)	1 gram (g) = 1,000 milligrams (mg)
Time	
1 year = 12 months	
1 year = 52 weeks	
1 week = 7 days	
1 day = 24 hours	
1 hour = 60 minutes	
1 minute = 60 seconds	

Metric Length Measurement

Convert to the units.

1) 5×10^4 mm = _____ cm

2) 0.4 m = _____ mm

3) 0.06 m = _____ cm

4) 1.2 km = _____ m

5) 8,000 mm = _____ m

6) 4,700 cm = _____ m

7) 4.5 m = _____ cm

8) 7×10^3 mm = _____ cm

9) 9×10^6 mm = _____ m

10) 2 km = _____ mm

11) 0.3 km = _____ m

12) 0.05 m = _____ cm

13) 4×10^4 m = _____ km

14) 6×10^7 m = _____ km

Customary Length Measurement

Convert to the units.

1) 20 ft = _____ in

2) 2.5 ft = _____ in

3) 5.6 yd = _____ ft

4) 0.4 yd = _____ ft

5) 9×10^{-1} yd = _____ in

6) 2 mi = _____ in

7) 18×10^3 in = _____ yd

8) 21.6 in = _____ yd

9) 6,160 yd = _____ mi

10) 28 yd = _____ in

11) 0.03 mi = _____ yd

12) 99×10^3 ft = _____ mi

13) 4.8 in = _____ ft

14) 42 yd = _____ feet

15) 0.72 in = _____ ft

16) 0.2 mi = _____ ft

SBAC Math Practice Grade 6

Metric Capacity Measurement

Convert the following measurements.

1) 60 l = _____ ml

2) 0.7 l = _____ ml

3) 2.8 l = _____ ml

4) 0.06 l = _____ ml

5) 22.5 l = _____ ml

6) 0.9 l = _____ ml

7) 6×10^6 ml = _____ l

8) 22×10^5 ml = _____ l

9) 112×10^2 ml = _____ l

10) 11,000 ml = _____ l

11) 57,800 ml = _____ l

12) 0.3×10^5 ml = _____ l

Customary Capacity Measurement

Convert the following measurements.

1) 1.5 gal = _____ qt.

2) 4.5 gal = _____ pt.

3) 0.5 gal = _____ c.

4) 18 pt. = _____ c

5) 12 c = _____ fl oz

6) 8.15 qt = _____ pt.

7) 0.08 qt = _____ c

8) 42 pt. = _____ c

9) 8×10^4 c = _____ gal

10) 256 pt. = _____ gal

11) 484 qt = _____ gal

12) 25.8 pt. = _____ qt

13) 7×10^3 c = _____ qt

14) 98.8 c = _____ pt.

15) 0.164 qt = _____ gal

16) 1,256 pt. = _____ qt

17) 23 gal = _____ pt.

18) 0.01 qt = _____ c

19) 800 c = _____ gal

20) 64.16 fl oz = _____ c

Metric Weight and Mass Measurement

Convert.

1) 0.8 kg = _____ g

2) 5.6 kg = _____ g

3) 2×10^{-4} kg = _____ g

4) 1.04 kg = _____ g

5) 44.8 kg = _____ g

6) 13.12 kg = _____ g

7) 0.072 kg = _____ g

8) 21×10^5 g = _____ kg

9) 15×10^6 g = _____ kg

10) 0.04×10^8 g = _____ kg

11) 17,400 g = _____ kg

12) 98×10^2 g = _____ kg

13) 5,400,000 g = _____ kg

14) 325×10^4 g = _____ kg

Customary Weight and Mass Measurement

Convert.

1) 24×10^4 lb. = _____ T

2) 0.32×10^5 lb. = _____ T

3) 190,000 lb. = _____ T

4) 2,800 lb. = _____ T

5) 0.35 lb. = _____ oz

6) 2.8 lb. = _____ oz

7) 0.05 lb. = _____ oz

8) 4 T = _____ lb.

9) 7×10^{-4} T = _____ lb.

10) 38×10^{-5} T = _____ lb.

11) 0.6 T = _____ lb.

12) 0.003 T = _____ oz

13) 0.015 T = _____ oz

14) 196.8 oz = _____ lb.

Unit of Measurements

Use the given ratios to convert the measuring units. If necessary, round the answers to three decimal digits.

1) Use $1 = \frac{1.6093 \text{km}}{1 \text{ mi}}$ and convert 6.02 miles to kilometers

   6.02 mi = _____

2) Use $1 = \frac{1.6093 \text{km}}{1 \text{ mi}}$ and convert 4.15 miles to kilometers

   4.15 mi = _____

3) Use $1 = \frac{1 \text{qt}}{0.946 \text{L}}$ and convert 6 liters to quarts

   6 L = _____

4) Use $1 = \frac{1 \text{qt}}{0.946 \text{L}}$ and convert 8 liters to quarts

   8 L = _____

5) Use $1 = \frac{1.6093 \text{km}}{1 \text{ mi}}$ and convert 5.06 miles to kilometers

   5.06 mi = _____

6) Use $1 = \frac{1.6093 \text{km}}{1 \text{ mi}}$ and convert 8.1 miles to kilometers

   8.1 mi = _____

7) Use $1 = \frac{1 \text{qt}}{0.946 \text{L}}$ and convert 5 liters to quarts

   5 L = _____

Temperature

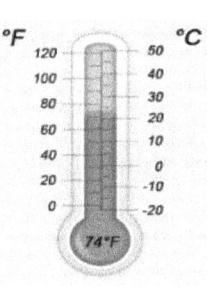

Convert Fahrenheit into Celsius.

1) 176°F = ___ °C

2) 113°F = ___ °C

3) 122°F = ___ °C

4) −13°F = ___ °C

5) 131°F = ___ °C

6) 136.4°F = ___ °C

7) 14°F = ___ °C

8) 149°F = ___ °C

9) 158°F = ___ °C

10) 167°F = ___ °C

11) 77°F = ___ °C

12) 86°F = ___ °C

Convert Celsius into Fahrenheit.

13) 85°C = ___ °F

14) 150°C = ___ °F

15) 83°C = ___ °F

16) 20°C = ___ °F

17) 5°C = ___ °F

18) −5°C = ___ °F

19) 0°C = ___ °F

20) 30°C = ___ °F

21) 90°C = ___ °F

22) 72°C = ___ °F

23) 38°C = ___ °F

24) 35°C = ___ °F

Time

Convert to the units.

1) 28 hr. = _____ min

2) 15 year = _____ week

3) 0.5 hr. = _____ sec

4) 8.5 min = _____ sec

5) 6×10^4 min = _____ hr

6) 1,095 day = _____ year

7) 2 year = _____ hr.

8) 42 day = _____ hr

9) 2 day = _____ min

10) 480 min = _____ hr

11) 28.5 year = _____ month

12) 12,600 sec = _____ min

13) 216 hr = _____ day

14) 15 weeks = _____ day

How much time has passed?

1) From 3:35 A.M. to 6:45 A.M.: ____ hours and ____ minutes.

2) From 2:30 A.M. to 7:15 A.M.: ____ hours and ____ minutes.

3) It's 6:20 P.M. What time was 3 hours ago? _____ O'clock

4) 4:15 A.M to 7:35 AM: _____ hours and _____ minutes.

5) 1:45 A.M to 4:20 AM: _____ hours and _____ minutes.

6) 9:00 A.M. to 10:05 AM. = _____ hour(s) and _____ minutes.

7) 10:35 A.M. to 3:05 PM. = _____ hour(s) and _____ minutes

8) 5:12 A.M. to 5:48 A.M. = _____ minutes

9) 8:08 A.M. to 8:45 A.M. = _____ minutes

Answers of Worksheets – Chapter 6

Metric length

1) 5,000 cm
2) 400 mm
3) 6 cm
4) 1,200 m
5) 8 m
6) 47 m
7) 450 cm
8) 700 cm
9) 9,000 m
10) 2,000,000 mm
11) 300 m
12) 5 cm
13) 40 km
14) 60,000 km

Customary Length

1) 240
2) 30
3) 16.8
4) 1.2
5) 32.4
6) 126,720
7) 500
8) 0.6
9) 3.5
10) 1,008
11) 52.8
12) 18.75
13) 0.4
14) 126
15) 0.06
16) 1,056

Metric Capacity

1) 60,000 ml
2) 700 ml
3) 2,800 ml
4) 60 ml
5) 22,500 ml
6) 900 ml
7) 6,000 ml
8) 2,200 ml
9) 11.2 ml
10) 11L
11) 57.8 L
12) 30 L

Customary Capacity

1) 6 qt
2) 36 pt.
3) 8 c
4) 36 c
5) 96 fl oz
6) 16.3 pt.
7) 0.32 c
8) 84 c
9) 5,000 gal
10) 32 gal
11) 121 gal
12) 12.9 qt
13) 1,750 qt
14) 49.4 pt.
15) 0.041 gal
16) 628 qt
17) 184 pt.
18) 0.04 c
19) 50 gal
20) 8.02 c

Metric Weight and Mass

1) 800 g
2) 5,600 g
3) 0.2 g
4) 1,040 g
5) 44,800 g
6) 13,120 g
7) 72 g
8) 2,100 kg
9) 15,000 kg
10) 4,000 kg
11) 17.4 kg
12) 9.8 kg
13) 5,400 kg
14) 3,250 kg

SBAC Math Practice Grade 6

Customary Weight and Mass

1) 120 T
2) 16 T
3) 95 T
4) 1.4 T
5) 5.6 oz
6) 44.8 oz
7) 0.8 oz
8) 8,000 lb.
9) 1.4 lb.
10) 0.76 lb.
11) 1,200 lb.
12) 96 oz
13) 480 oz
14) 12.3 lb

Unit of measurements

1) 9.688 km
2) 6.679 km
3) 6.342 qt
4) 8.457 qt
5) 8.143 km
6) 13.035 km
7) 5.285 qt

Temperature

1) 80°C
2) 45°C
3) 50°C
4) −25°C
5) 55°C
6) 58°C
7) −10°C
8) 65°C
9) 70°C
10) 75°C
11) 25°C
12) 30°C
13) 185°F
14) 302°F
15) 181.4°F
16) 68°F
17) 41°F
18) 23°F
19) 32°F
20) 86°F
21) 194°F
22) 161.6°F
23) 100.4°F
24) 95°F

Time - Convert

1) 1,680 min
2) 780 weeks
3) 1,800 sec
4) 510 sec
5) 1,000 hr
6) 3 year
7) 17,520 hr
8) 1,008 hr
9) 2,880 min
10) 8 hr
11) 342 months
12) 210 min
13) 9 days
14) 105 days

Time - Gap

1) 3:10
2) 4:45
3) 3:20 P.M.
4) 3:20
5) 2:35
6) 1:05
7) 4:30
8) 36 minutes
9) 37 minutes

Chapter 7:
Algebraic Expressions

Find a Rule

Complete the output.

1- **Rule: the output is** $x + 25$

Input	x	8	15	20	38	40
Output	y					

2- **Rule: the output is** $x \times 18$

Input	x	3	7	10	11	15
Output	y					

3- **Rule: the output is** $x \div 7$

Input	x	126	147	105	280	455
Output	y					

Find a rule to write an expression.

4- **Rule:** _____

Input	x	11	13	15	20
Output	y	55	65	75	100

5- **Rule:** _____

Input	x	10	28	32	46
Output	y	14	32	36	50

6- **Rule:** _____

Input	x	84	132	180	252
Output	y	14	22	30	42

Variables and Expressions

Write a verbal expression for each algebraic expression.

1) $2a - 4b$

2) $8c^2 + 2d$

3) $x - 8$

4) $\frac{80}{15}$

5) $a^2 + b^3$

6) $2x + 4$

7) $x^2 - 10y + 8$

8) $x^3 + 9y^2 - 4$

9) $\frac{1}{3}x + \frac{3}{4}y - 6$

10) $\frac{1}{5}(x + 8) - 10y$

Write an algebraic expression for each verbal expression.

11) 9 less than h

12) The product of 9 and b

13) The 26 divided by k

14) The product of 5 and the third power of x

15) 10 more than h to the fifth power

16) 20 more than twice d

17) One fourth the square of b

18) The difference of 23 and 4 times a number

19) 60 more than the cube of a number

20) Three-quarters the cube of a number

Translate Phrases

Write an algebraic expression for each phrase.

1) A number increased by sixty–one.

2) The sum of twenty and 2 times a number

3) The difference between fifty–seven and a number.

4) The quotient of twenty-two and a number.

5) Twice a number decreased by 50.

6) four times the sum of a number and − 20.

7) A number divided by − 12.

8) The quotient of 49 and the product of a number and − 12.

9) ten subtracted from 2 times a number.

10) The difference of eight and a number.

Distributive Property

Multiply using the distributive property.

1) $5(x + 8) =$ _____

2) $2(x + 9) =$ _____

3) $(x + 4)6 =$ _____

4) $3(x + 5) =$ _____

5) $9(x + 7) =$ _____

6) $12(x + 3) =$ _____

7) $11(x + 2) =$ _____

8) $8(x + 9) =$ _____

9) $9(x + 9) =$ _____

10) $(x + 7)7 =$ _____

11) $(x + 10)5 =$ _____

12) $2(x + 13) =$ _____

13) $3(5x - 7) =$ _____

14) $4(6x - 5) =$ _____

15) $6(5x - 4) =$ _____

16) $(3x - 9)2 =$ _____

17) $(9x - 3)6 =$ _____

18) $(8x - 4)9 =$ _____

19) $5(7x - 6) =$ _____

20) $(-3)(8x - 8) =$ _____

21) $(-4)(x - 11) =$ _____

22) $(-9)(5x - 2) =$ _____

23) $(6x + 5)(-8) =$ _____

24) $(x + 8)(-11) =$ _____

Distributive and Simplifying Expressions

Simplify each expression.

1) $6x + 2 - 8 =$

2) $-(-4 - 5x) =$

3) $(-3x + 4)(-2) =$

4) $(-2x)(x + 3) =$

5) $-2x + x^2 + 4x^2 =$

6) $7y + 7x + 8y - 5x =$

7) $-3x + 3y + 14x - 9y =$

8) $-2x - 5 + 8x + \frac{16}{4} =$

9) $5 - 8(x - 2) =$

10) $-5 - 5x + 3x =$

11) $(x - 3y)2 + 4y =$

12) $2.5x^2 \times (-5x) =$

13) $-4 - 2x^2 + 6x^2 =$

14) $8 + 14x^2 + 4 =$

15) $4(-2x - 7) + 10 =$

16) $(-x)(-2 + 3x) - x(7 + x) =$

17) $-3(6 + 12) - 3x + 5x =$

18) $-4(5 - 12x - 3x) =$

19) $3(-2x - 6) =$

20) $9 + 7x - 9 =$

21) $x(-2x + 8) =$

22) $5xy + 4x - 3y + x + 2y =$

23) $3(-x - 7) + 9 =$

24) $(-3x - 4) + 7 =$

25) $3x + 4y - 5 + 1 =$

26) $(-2 + 3x) - 3x(1 + 2x) =$

27) $(-3)(-3x - 3y) =$

28) $4(-x - 2) + 5 =$

Factoring Expressions

Factor the common factor out of each expression.

1) $12x - 6 =$

2) $5x - 15 =$

3) $\frac{45}{15}x - 15 =$

4) $7b - 28 =$

5) $4a^2 - 24a =$

6) $2xy - 10y =$

7) $5x^2y + 15x =$

8) $a^2 - 8a + 7ab =$

9) $2a^2 + 2ab =$

10) $4x + 20 =$

11) $24x - 36xy =$

12) $8x - 6 =$

13) $\frac{1}{4}x - \frac{3}{4}y =$

14) $7xy - \frac{14}{3}x =$

15) $3ab + 9c =$

16) $\frac{1}{3}x - \frac{4}{3} =$

17) $10x - 15xy =$

18) $x^2 + 8x =$

19) $4x^2 - 12y =$

20) $4x^3 + 3xy + x^2 =$

21) $21x - 14 =$

22) $20b - 60c + 20d =$

23) $24ab - 8ac =$

24) $ax - ay - 3x + 3y =$

25) $3ax + 4a + 9x + 12 =$

26) $x^2 - 10x =$

27) $9x^3 - 18x^2 =$

28) $5x^2 - 70xy =$

Evaluate One Variable Expressions

Evaluate each using the values given.

1) $x + 4x, x = 3$

2) $5(-6 + 3x), x = 1$

3) $4x + 7x, x = -3$

4) $5(2 - x) + 5, x = 3$

5) $6x + 4x - 10, x = 2$

6) $5x + 11x + 12, x = -1$

7) $5x - 2x - 4, x = 5$

8) $\frac{3(5x+8)}{9}, x = 2$

9) $2x - 85, x = 32$

10) $\frac{x}{18}, x = 108$

11) $7(3 + 2x) - 33, x = 5$

12) $7(x + 3) - 23, x = 4$

13) $\frac{x+(-6)}{-3}, x = -6$

14) $8(6 - 3x) + 5, x = 2$

15) $-11 - \frac{x}{5} + 3x, x = 10$

16) $5x + 11x, x = 1$

17) $-12x + 3(5 + 3x), x = -7$

18) $x + 11x, x = 0.5$

19) $\frac{(2x-2)}{6}, x = 13$

20) $3(-1 - 2x), x = 5$

21) $5x - (5 - x), x = 3$

22) $\left(-\frac{15}{x}\right) + 2 + x, x = 5$

23) $-\frac{x \times 5}{x}, x = 5$

24) $2(-1 - 3x), x = 2$

25) $2x^2 + 7x, x = 1$

26) $2(3x + 1) - 4(x - 5), x = 3$

27) $-6x - 4, x = -5$

28) $7x + 2x, x = 3$

Evaluate Two Variable Expressions

Evaluate the expressions.

1) $x + 4y, \ x = 5, y = 2$

2) $(-2)(-3x - 2y), \ x = 1, y = 2$

3) $4x + 2y, \ x = 10, y = 5$

4) $\frac{x-4}{y+1}, \ x = 8, y = 3$

5) $\frac{a}{4} - 6b, \ a = 32, b = 4$

6) $3x - 4(y - 8), \ x = 5, y = 3$

7) $3x + 2y - 10, \ x = 2, y = 10$

8) $-3x + 10 + 8y - 5, \ x = 2, y = 1$

9) $yx \div 3, \ x = 9, y = 9$

10) $a - b \div 3, \ a = 3, b = 12$

11) $6(x - y), \ x = 7, y = 4$

12) $5x - 4y, \ x = 5, y = 8$

13) $\frac{10}{a} + 3b, \ a = 5, b = 4$

14) $2x^2 + 4xy, \ x = 3, y = 5$

15) $8 - \frac{xy}{10} + y, \ x = 6, y = 5$

16) $7(3x - y), \ x = 7, y = -9$

17) $5x^2 - 3y^2, \ x = -1, y = 2$

18) $3x + \frac{y}{4}, \ x = 6, y = 16$

19) $4(4x - 2y), \ x = 3, y = 5$

20) $4x(y - \frac{1}{2}), \ x = 5, y = 4$

21) $5(x^2 - 2y), \ x = 3, y = 2$

22) $5xy, \ x = 2, y = 8$

23) $\frac{1}{3}y^3 \left(y - \frac{1}{4}x\right), \ x = -4, y = 3$

24) $-3(x - 5y) - 2x, \ x = 4, y = 2$

25) $-2x + \frac{1}{6}xy, \ x = 3, y = 6$

26) $x^2 + xy^2, \ x = 5, y = 7$

27) $x - 2y + 8, \ x = 9, y = 6$

28) $\frac{xy}{2x+y}, \ x = 5, y =$

SBAC Math Practice Grade 6

Finding Distance of Two Points

Find the distance between each pair of points.

1) $(2, 1), (-1, -3)$

2) $(-4, -2), (4, 4)$

3) $(-3, 0), (15, 24)$

4) $(-4, -1), (1, 11)$

5) $(3, -2), (-6, -14)$

6) $(-6, 0), (-2, 3)$

7) $(3, 2), (11, 17)$

8) $(-6, -10), (6, -1)$

9) $(5, 9), (-11, -3)$

10) $(6, -2), (2, -6)$

11) $(3, 0), (18, 36)$

12) $(8, 4), (3, -8)$

13) $(4, 2), (-5, -10)$

14) $(-8, 10), (4, 40)$

15) $(8, 4), (-10, -20)$

16) $(-8, -2), (16, 8)$

17) $(3, 5), (-5, -10)$

18) $(-10, 20), (35, 45)$

Find the midpoint of the line segment with the given endpoints.

1) $(-2, -2), (4, 2)$

2) $(10, 4), (-2, 4)$

3) $(12, -2), (4, 10)$

4) $(-6, -5), (2, 1)$

5) $(3, -2), (5, -2)$

6) $(-10, -4), (6, -2)$

7) $(4, 1), (-4, 9)$

8) $(-5, 6), (-5, 2)$

9) $(-8, 8), (4, -2)$

10) $(1, 7), (5, -1)$

11) $(-9, 5), (5, 3)$

12) $(7, 10), (-3, -6)$

13) $(-8, 14), (-8, 2)$

14) $(16, 7), (6, -3)$

15) $(5, 6), (-3, 4)$

16) $(-9, -1), (-5, 7)$

17) $(17, 9), (5, 11)$

18) $(-8, -11), (18, -1)$

Answer key Chapter 7

Find a Rule

1)

Input	x	8	15	20	38	40
Output	y	33	40	45	63	65

2)

Input	x	3	7	10	11	15
Output	y	54	126	180	198	270

3)

Input	x	126	147	105	280	455
Output	y	18	21	15	40	65

4) y = 5x   5) y = x + 4   6) y = x ÷ 6

Variables and Expressions

1) 2 times a minus 4 times b

2) 8 times c squared plus 2 times d

3) a number minus 8

4) the quotient of 80 and 15

5) a squared plus b cubed

6) the product of 2 and x plus 4

7) x squared minus the product of 10 and y plus 8

8) x cubed plus the product of 9 and y squared minus 4

9) the sum of one–thirds of x and three–quarters of y, minus 6

10) one–fifth of the sum of x and 8 minus the product of 10 and y

11) 9 < h

12) 9b

13) $\frac{26}{K}$

14) $5x^3$

15) $10 > h^5$

16) 2d < 20

17) $\frac{1}{4}b^2$

18) 23 − 4a

19) $60 > a^3$

20) $\frac{3}{4}x^3$

Translate Phrases

1) x + 61

2) 20 + 2x

3) 57 − x

4) $\frac{22}{x}$

5) 2x − 50

6) 4(x + (−20))

7) $\frac{x}{-12}$

8) $\frac{49}{-12x}$

9) 2x − 10

10) 8 − x

Distributive Property

1) 5x + 40

2) 2x + 18

3) 6x + 24

4) 3x + 15

5) 9x + 63

6) 12x + 36

SBAC Math Practice Grade 6

7) $11x + 22$
8) $8x + 72$
9) $9x + 81$
10) $7x + 49$
11) $5x + 50$
12) $2x + 26$

13) $15x - 21$
14) $24x - 20$
15) $30x - 24$
16) $6x - 18$
17) $54x - 18$
18) $72x - 36$

19) $35x - 30$
20) $-24x + 24$
21) $-4x + 44$
22) $-45x + 18$
23) $-48x - 40$
24) $-11x - 88$

Distributive and Simplifying Expressions

1) $6x - 6$
2) $4 + 5x$
3) $6x - 8$
4) $-2x^2 - 6x$
5) $5x^2 - 2x$
6) $2x + 15y$
7) $11x - 6y$
8) $6x - 1$
9) $-8x + 21$
10) $-2x - 5$

11) $2x - 2y$
12) $-12.5x^3$
13) $4x^2 - 4$
14) $14x^2 + 12$
15) $-8x - 18$
16) $-4x^2 - 5x$
17) $2x - 54$
18) $60x - 20$
19) $-6x - 18$
20) $7x$

21) $-2x^2 + 8x$
22) $5x + y + 5xy$
23) $-3x - 12$
24) $-3x + 3$
25) $3x + 4y - 4$
26) $-6x^2 - 2$
27) $9x + 9y$
28) $-4x - 3$

Factoring Expressions

1) $3(4x - 2)$
2) $5(x - 3)$
3) $3(x - 5)$
4) $7(b - 4)$
5) $4a(a - 6)$
6) $2y(x - 5)$
7) $5x(xy + 3)$
8) $a(a - 8 + 7b)$
9) $2a(a + b)$
10) $4(x + 5)$

11) $12x(2 - 3y)$
12) $2(4x - 3)$
13) $\frac{1}{4}(x - 3y)$
14) $7x(y - \frac{2}{3})$
15) $3(ab + 3c)$
16) $\frac{1}{3}(x - 4)$
17) $5x(2 - 3y)$
18) $x(x + 8)$
19) $4(x^2 - 3y)$

20) $x(4x^2 + 3y + x)$
21) $7(3x - 2)$
22) $20(b - 3c + d)$
23) $8a(3b - c)$
24) $(x - y)(a - 3)$
25) $(3x + 4)(a + 3)$
26) $x(x - 10)$
27) $9x^2(x - 2)$
28) $5x(x - 14y)$

Evaluate One Variable Expressions

1) 15
2) -15
3) -33
4) 0
5) 10
6) -4
7) 11
8) 6

9) −21
10) 6
11) 58
12) 26
13) 4
14) 5
15) 17
16) 16
17) 36
18) 6
19) 4
20) −33
21) 13
22) 4
23) −5
24) −14
25) 9
26) 28
27) 26
28) 27

Evaluate Two Variable Expressions

1) 13
2) 14
3) 50
4) 1
5) −16
6) 35
7) 16
8) 7
9) 27
10) 3
11) 18
12) 20
13) 17
14) 78
15) 10
16) 210
17) −7
18) 22
19) 8
20) 70
21) 25
22) 80
23) 36
24) 10
25) −3
26) 270
27) 5
28) $\frac{10}{7}$

Finding Distance of Two Points

1) 5
2) 10
3) 30
4) 13
5) 15
6) 5
7) 17
8) 15
9) 20
10) $4\sqrt{2}$
11) 39
12) 13
13) 15
14) $6\sqrt{29}$
15) 30
16) 26
17) 17
18) $5\sqrt{106}$

Finding Midpoint

1) (1, 0)
2) (4, 4)
3) (8, 4)
4) (−2, −2)
5) (4, −2)
6) (−2, −3)
7) (0, 5)
8) (−5, 4)
9) (−2, 3)
10) (3, 3)
11) (−2, 4)
12) (2, 2)
13) (−8, 8)
14) (11, 2)
15) (1, 5)
16) (−7, 3)
17) (11, 10)
18) (5, −6)

Chapter 8:
Equations

Graphing Linear Equation

Sketch the graph of each line.

1) $y = 2x - 5$ 2) $y = -2x + 3$ 3) $x - y = 0$

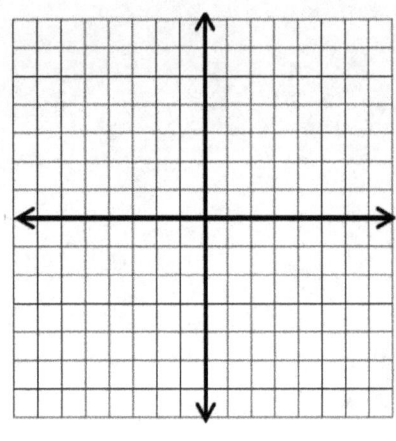 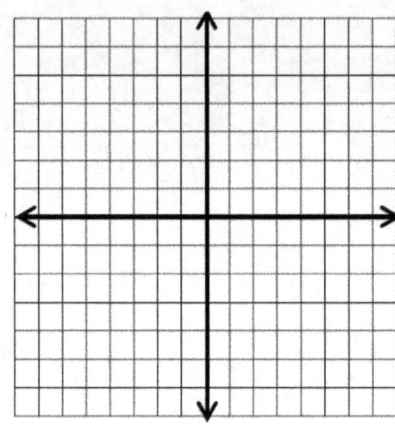 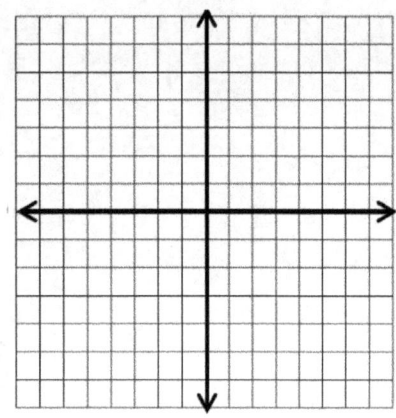

4) $x + y = 3$ 5) $5x + 3y = -2$ 6) $y - 3x + 2 = 0$

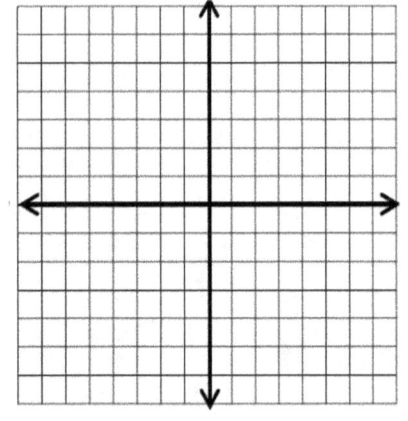 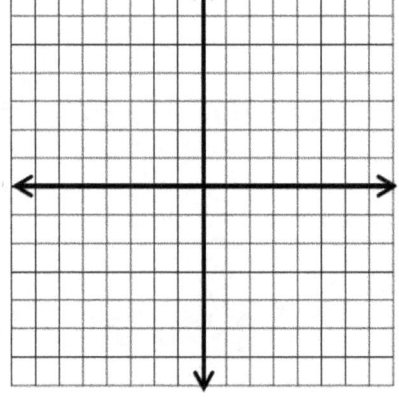 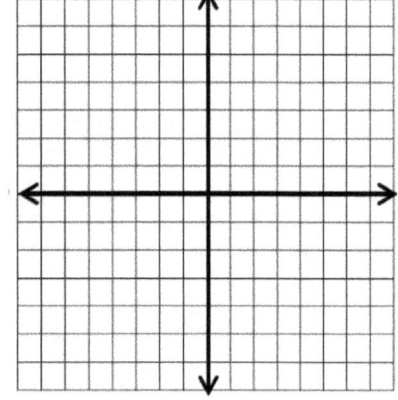

One Step Equations

Solve each equation.

1) $44 = (-12) + x$

2) $8x = (-64)$

3) $(-72) = (-8x)$

4) $(-5) = 3 + x$

5) $4 + \frac{x}{2} = (-3)$

6) $8x = (-104)$

7) $62 = x - 13$

8) $\frac{x}{3} = (-15)$

9) $x + 112 = 154$

10) $x - \frac{1}{3} = \frac{2}{3}$

11) $(-24) = x - 32$

12) $(-3x) = 39$

13) $(-169) = (13x)$

14) $-4x + 42 = 50$

15) $5x + 3 = 38$

16) $80 = (-8x)$

17) $3x + 7 = 19$

18) $24x = 144$

19) $x - 18 = 15$

20) $0.9x = 4.5$

21) $4x = 84$

22) $2x + 2.98 = 66.98$

23) $x + 9 = 6$

24) $x + 14 = 6$

25) $9x + 41 = 5$

26) $\frac{1}{4}x + 30 = 12$

Two Steps Equations

Solve each equation.

1) $6(3 + x) = 42$

2) $(-7)(x - 2) = 56$

3) $(-8)(3x - 4) = (-16)$

4) $5(2 + x) = -15$

5) $19(3x + 11) = 38$

6) $4(2x + 2) = 24$

7) $5(8 + 3x) = (-20)$

8) $(-5)(5x - 3) = 40$

9) $2x + 12 = 16$

10) $\frac{4x - 5}{5} = 3$

11) $(-3) = \frac{x + 4}{7}$

12) $80 = (-8)(x - 3)$

13) $\frac{x}{3} + 7 = 19$

14) $\frac{1}{4} = \frac{1}{2} + \frac{x}{4}$

15) $\frac{11 + x}{5} = (-6)$

16) $(-3)(10 + 5x) = (-15)$

17) $(-3x) + 12 = 24$

18) $\frac{x + 5}{5} = -5$

19) $\frac{x + 23}{8} = 3$

20) $(-4) + \frac{x}{2} = (-14)$

21) $-5 = \frac{x + 7}{8}$

22) $\frac{9x - 3}{6} = 4$

23) $\frac{2x - 12}{8} = 6$

24) $40 = (-5)(x - 8)$

Multi Steps Equations

Solve each equation.

1) $2 - (4 - 5x) = 3$

2) $-15 = -(4x + 7)$

3) $6x - 18 = (-2x) + 6$

4) $-32 = (-5x) - 11x$

5) $3(2 + 3x) + 3x = -30$

6) $5x - 18 = 2 + 2x - 7 + 2x$

7) $12 - 6x = (-36) - 3x + 3x$

8) $16 - 4x - 4x = 8 - 4x$

9) $8 + 7x + x = (-12) + 3x$

10) $(-3x) - 3(-2 + 4x) = 366$

11) $20 = (-200x) - 5 + 5$

12) $61 = 5x - 23 + 7x$

13) $7(4 + 2x) = 140$

14) $-60 = (-7x) - 13x$

15) $2(4x + 5) = -2(x + 4) - 22$

16) $11x - 17 = 6x + 8$

17) $9 = -3(x - 8)$

18) $(-6) - 8x = 6(1 + 2x)$

19) $x + 3 = -2(9 + 3x)$

20) $10 = 4 - 5x - 9$

21) $-15 - 9x - 3x = 12 - 3x$

22) $-23 - 3x + 5x = 27 - 23x$

23) $19 - 6x - 9x = -5 - 9x$

24) $15x - 18 = 6x + 9$

Answer key Chapter 8

Graphing Lines Using Line Equation

1) $y = 2x - 5$

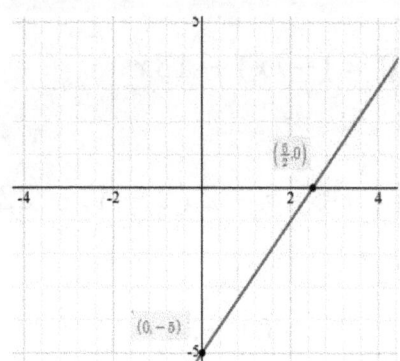

2) $y = -2x + 3$

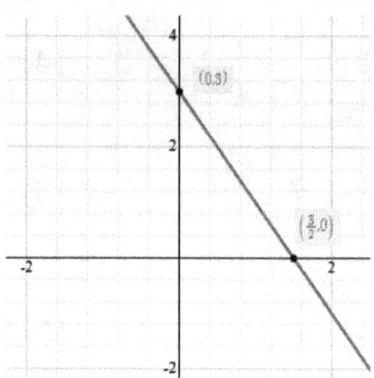

3) $x - y = 0$

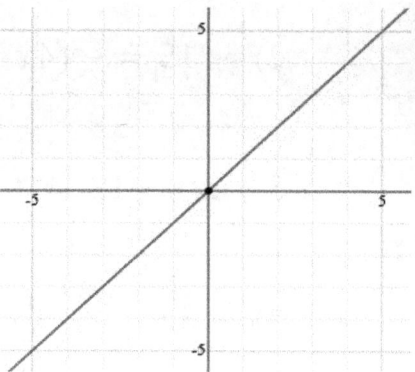

4) $x + y = 3$

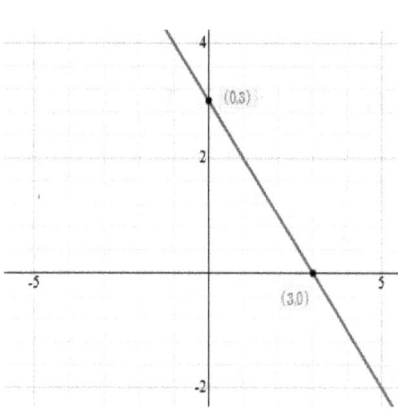

5) $5x + 3y = -2$

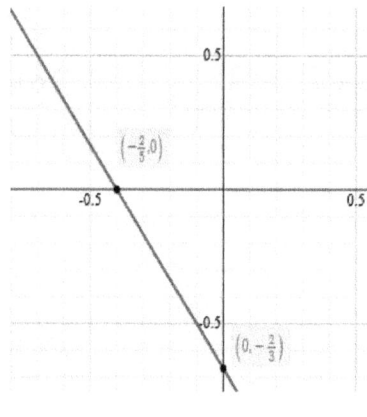

6) $y - 3x + 2 = 0$

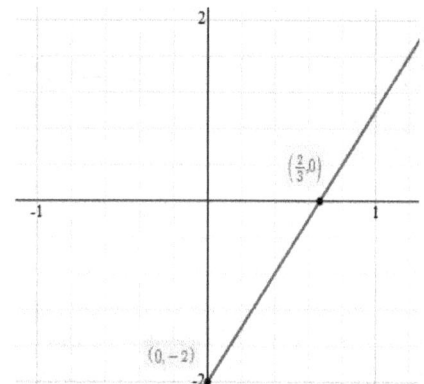

One Step Equations

1) $x = 56$
2) $x = -8$
3) $x = 9$
4) $x = -8$
5) $x = -14$
6) $x = -13$
7) $x = 75$
8) $x = -45$
9) $x = 42$
10) $x = 1$
11) $x = 8$
12) $x = -13$
13) $x = -13$
14) $x = -2$
15) $x = 7$
16) $x = -10$
17) $x = 4$
18) $x = 6$
19) $x = 33$
20) $x = 5$
21) $x = 21$

22) x = 32
23) $x = -3$
24)
25)
26)

Two Steps Equations

1) $x = 4$
2) $x = -6$
3) $x = 2$
4) $x = -5$
5) $x = -3$
6) $x = 2$
7) $x = -4$
8) $x = -1$
9) $x = 2$
10) $x = 5$
11) $x = -25$
12) $x = -7$
13) $x = 36$
14) $x = -1$
15) $x = -41$
16) $x = -1$
17) $x = -4$
18) $x = -30$
19) $x = 1$
20) $x = -20$
21) $x = -47$
22) $x = 3$
23) $x = 30$
24) $x = 0$

Multi Steps Equations

1) $x = 1$
2) $x = 2$
3) $x = 3$
4) $x = 2$
5) $x = -3$
6) $x = 13$
7) $x = 8$
8) $x = 2$
9) $x = -4$
10) $x = -24$
11) $x = -0.1$
12) $x = 7$
13) $x = 8$
14) $x = 3$
15) $x = -4$
16) $x = 5$
17) $x = 5$
18) $x = -3/5$
19) $x = -3$
20) $x = -3$
21) $x = -3$
22) $x = 2$
23) $x = 4$
24) $x = 3$

Chapter 9:

Inequality

Graphing Linear Inequalities

Sketch the graph of each linear inequality.

1) $y > 2x - 3$
2) $y < x + 3$
3) $y \leq -3x - 8$

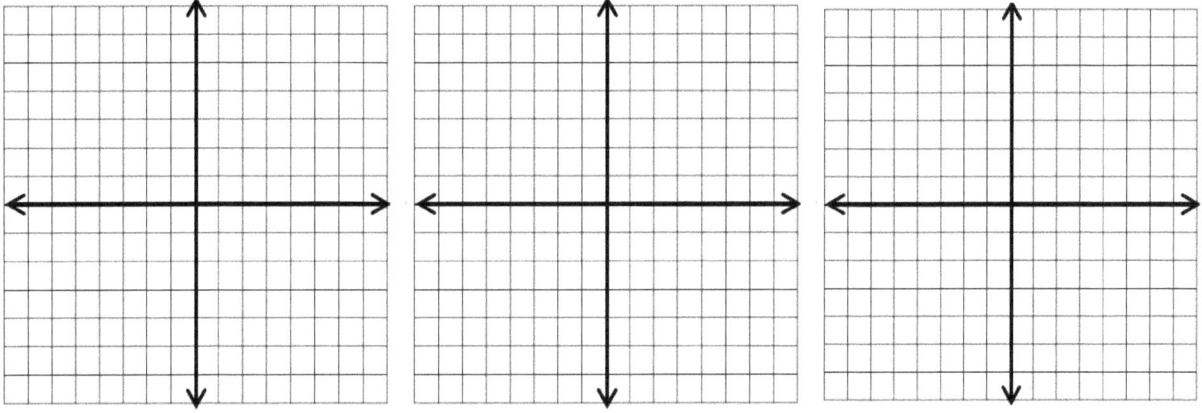

4) $3y \geq 6 + 3x$
5) $-3y < x - 12$
6) $2y \geq -8x + 4$

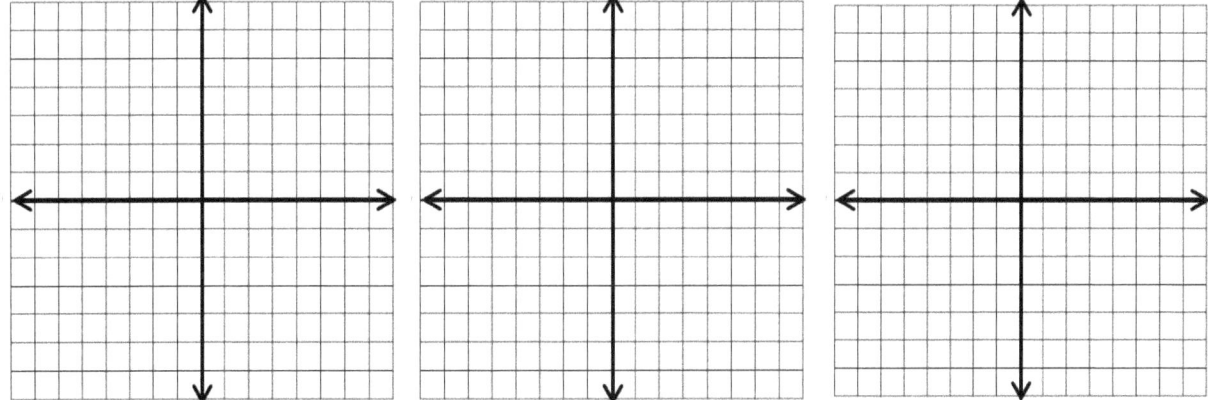

One Step Inequality

Solve each inequality.

1) $7x < 14$

2) $x + 7 \geq -8$

3) $x - 1 \leq 9$

4) $-2x + 4 > -10$

5) $x + 18 \geq -6$

6) $x + 9 \geq 5$

7) $x - \frac{1}{3} \leq 5$

8) $-7x < 42$

9) $-x + 8 > -3$

10) $\frac{x}{3} + 3 > -9$

11) $-x + 8 > -4$

12) $x - 14 \leq 18$

13) $-x - 5 \leq -7$

14) $x + 26 \geq -13$

15) $x + \frac{1}{3} \geq -\frac{2}{3}$

16) $x + 6 \geq -14$

17) $x - 42 \leq -48$

18) $x - 5 \leq 4$

19) $-x + 5 > -6$

20) $x + 6 \geq -12$

21) $8x + 6 \leq 22$

22) $4x - 3 \geq 9$

23) $3x - 5 < 22$

24) $6x - 8 \leq 40$

Two Steps Inequality

Solve each inequality

1) $2x - 3 \leq 7$

2) $3x - 4 \leq 8$

3) $\frac{-1}{4}x + \frac{x}{2} \leq \frac{1}{8}$

4) $5x + 10 \geq 30$

5) $4x - 7 \geq 9$

6) $3x - 5 \leq 16$

7) $8x - 2 \leq 14$

8) $9x + 5 \leq 23$

9) $2x + 10 > 32$

10) $\frac{x}{8} + 2 \leq 4$

11) $3x + 4 \geq 37$

12) $3x - 8 < 10$

13) $6 \geq \frac{x+7}{2}$

14) $3x + 9 < 48$

15) $\frac{4+x}{5} \geq 3$

16) $16 + 4x < 36$

17) $16 > 6x - 8$

18) $5 + \frac{x}{3} < 6$

19) $-4 + 4x > 24$

20) $5 + \frac{x}{7} < 3$

Multi Steps Inequality

Solve the inequalities.

1) $4x - 6 < 5x - 9$

2) $\frac{4x+5}{3} \leq x$

3) $7x - 5 > 3x + 15$

4) $-3x > -6x + 4$

5) $3 + \frac{x}{2} < \frac{x}{4}$

6) $\frac{4x-6}{8} > x$

7) $4x - 20 + 4 > 6x - 8$

8) $x - 8 > 11 + 3(x + 5)$

9) $\frac{x}{3} + 2 > x$

10) $-7x + 8 \geq -6(4x - 8) - 8x$

11) $7x - 4 \leq 8x + 9$

12) $\frac{2x-7}{5} > 2$

13) $8(x + 2) < 6x + 10$

14) $-8x + 12 \leq 4(x - 9)$

15) $\frac{5x-6}{3} > 3x + 2$

16) $2(x - 8) + 10 \geq 4x - 2$

17) $\frac{-5x+7}{6} > 5x$

18) $-3x - 4 > -7x$

19) $\frac{1}{4}x - 12 > \frac{1}{8}x - 19$

20) $-4(x - 9) \leq 5x$

Answer key Chapter 9

Graphing Linear Inequalities

1) $y > 2x - 3$
2) $y < x + 3$
3) $y \leq -3x - 8$

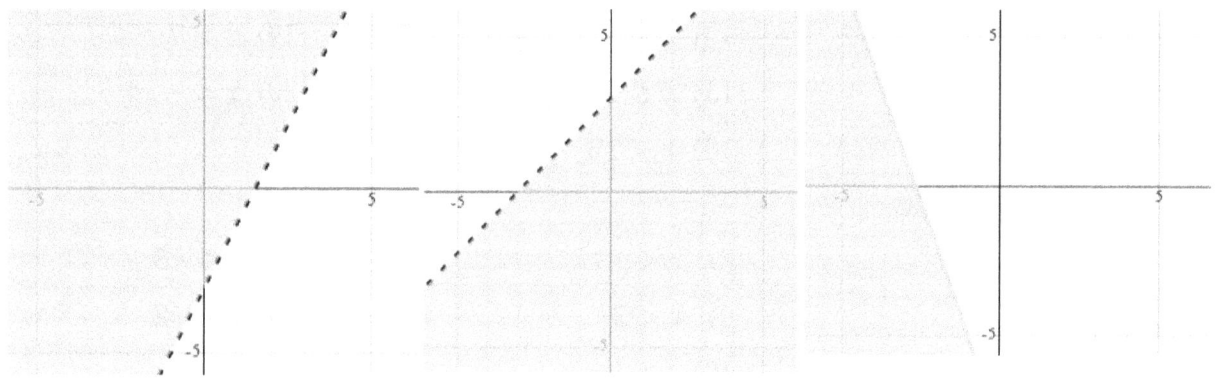

4) $3y \geq 6 + 3x$
5) $-3y < x - 12$
6) $2y \geq -8x + 4$

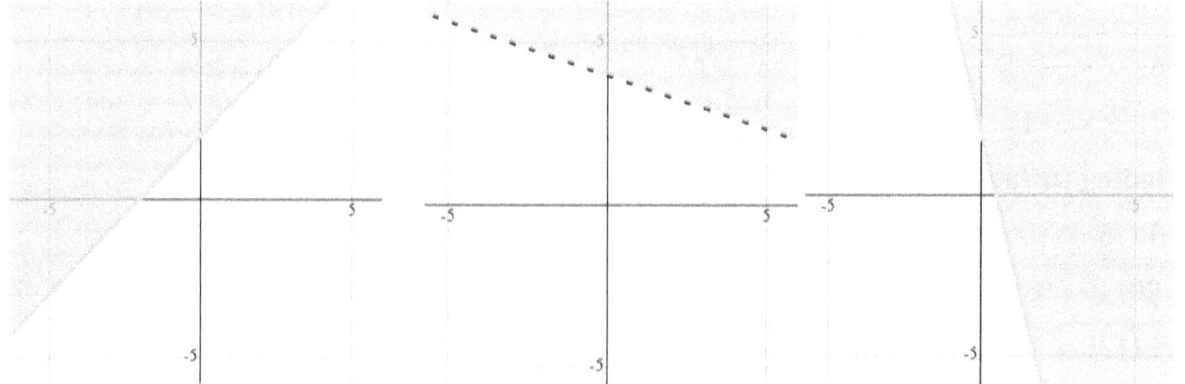

One Step Inequality

1) $x < 2$
2) $x \geq -15$
3) $x \leq 10$
4) $x < 7$
5) $x \geq -24$
6) $x \geq -4$
7) $x \leq \frac{16}{3}$
8) $x > -6$
9) $x < 11$
10) $x > -36$
11) $x < 12$
12) $x \leq 32$
13) $x \geq 2$
14) $x \geq -39$
15) $x \geq -1$
16) $x \geq -20$
17) $x \leq -6$
18) $x \leq 9$
19) $x < 11$
20) $x \geq -18$
21) $x \leq 2$
22) $x \geq 3$
23) $x < 9$
24) $x \leq 8$

Two Steps Inequality

1) $x \leq 5$
2) $x \leq 4$
3) $x \leq 0.5$
4) $x \geq 4$
5) $x \geq 4$
6) $x \leq 7$
7) $x \leq 2$

8) $x \leq 2$
9) $x > 11$
10) $x \leq 16$
11) $x \geq 11$
12) $x < 6$
13) $x \leq 5$
14) $x < 13$

15) $x \geq 11$
16) $x < 5$
17) $x < 4$
18) $x < 3$
19) $x > 7$
20) $x < -14$

Multi Steps Inequality

1) $x > 3$
2) $x \leq -5$
3) $x > 5$
4) $x > \frac{4}{3}$
5) $x < -12$
6) $x < -1.5$
7) $x < -4$

8) $x < -17$
9) $x < 3$
10) $x \geq 1.6$
11) $x \geq -13$
12) $x > 8.5$
13) $x < -3$
14) $x \geq 4$

15) $x < -3$
16) $x \leq -2$
17) $x < \frac{1}{5}$
18) $x > 1$
19) $x > -56$
20) $x \geq 4$

Finding Distance of Two Points

19) 5
20) 10
21) 30
22) 13
23) 15
24) 5

25) 17
26) 15
27) 20
28) $4\sqrt{2}$
29) 39
30) 13

31) 15
32) $6\sqrt{29}$
33) 30
34) 26
35) 17
36) $5\sqrt{106}$

Finding Midpoint

19) $(1, 0)$
20) $(4, 4)$
21) $(8, 4)$
22) $(-2, -2)$
23) $(4, -2)$
24) $(-2, -3)$

25) $(0, 5)$
26) $(-5, 4)$
27) $(-2, 3)$
28) $(3, 3)$
29) $(-2, 4)$
30) $(2, 2)$

31) $(-8, 8)$
32) $(11, 2)$
33) $(1, 5)$
34) $(-7, 3)$
35) $(11, 10)$
36) $(5, -6)$

Chapter 10:

Geometry

Area and Perimeter of Square

Find the perimeter and area of each squares.

1)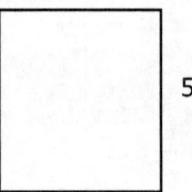

Perimeter:_____:

Area:_____:

2)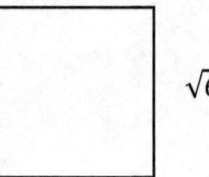

Perimeter:_____:

Area:_____:

3)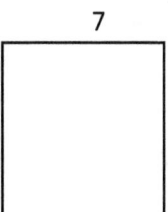

Perimeter:_____:

Area:_____:

4)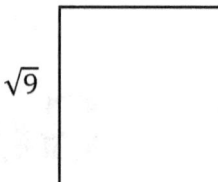

Perimeter:_____:

Area:_____:

5)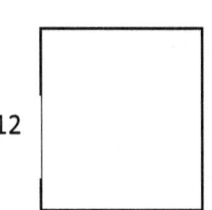

Perimeter:_____:

Area:_____:

6)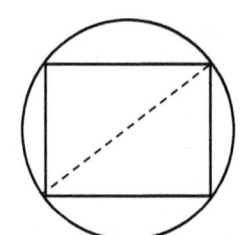

Perimeter of Square:_____:

Area of Square:_____:

Area and Perimeter of Rectangle

Find the perimeter and area of each rectangle.

1)

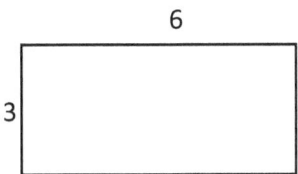

Perimeter: ....:

Area: ....:

2)

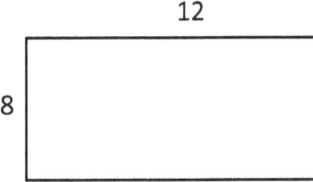

Perimeter: ....:

Area: ....:

3)

15

10

Perimeter: ....:

Area: ....:

4)

7

2.5

Perimeter: ....:

Area: ....:

5)

$1\frac{2}{5}$

$\frac{5}{7}$

Perimeter: ....:

Area: ....:

6)

5

2

Perimeter: ....:

Area: ....:

Area and Perimeter of Triangle

Find the perimeter and area of each triangle.

1)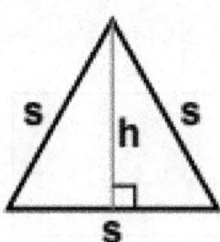

Perimeter:_____:

Area:_____:

2)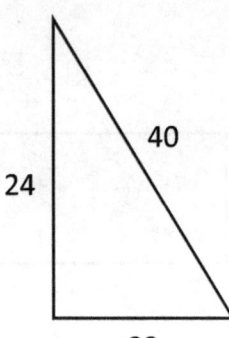

Perimeter:_____:

Area:_____:

3)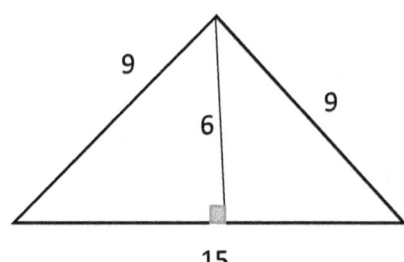

Perimeter:_____:

Area _____:

4)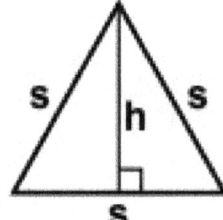

s=8

h=6

Perimeter:_____:

Area:_____:

5)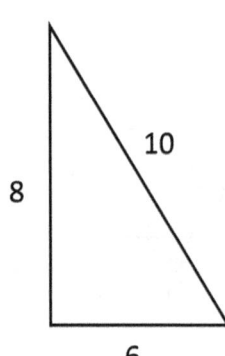

Perimeter:_____:

Area:_____:

6)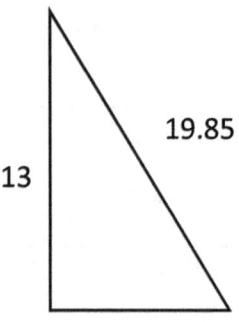

Perimeter:_____:

Area:_____:

Area and Perimeter of Trapezoid

Find the perimeter and area of each trapezoid.

1)

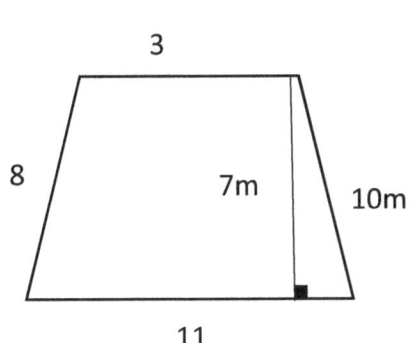

Perimeter: _____:

Area: _____:

2)

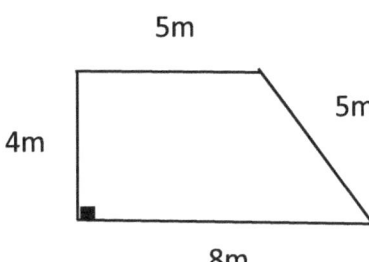

Perimeter: _____:

Area: _____:

3)

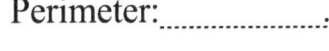

Perimeter: _____:

Area _____:

4)

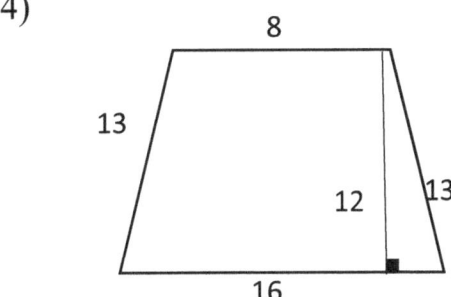

Perimeter: _____:

Area: _____:

5)

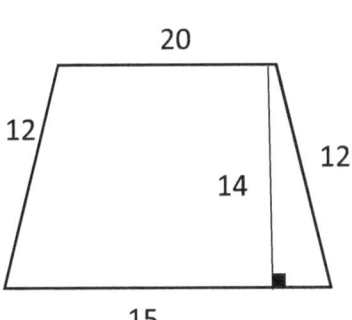

Perimeter: _____:

Area: _____:

6)

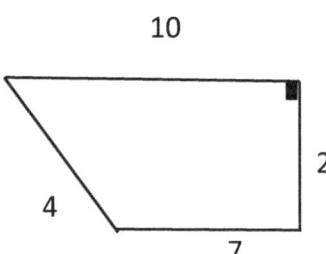

Perimeter: _____:

Area: _____:

Area and Perimeter of Parallelogram

Find the perimeter and area of each parallelogram.

1)

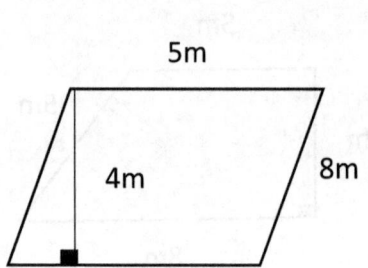

Perimeter:_____.

Area:_____.

2)

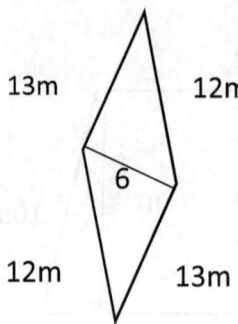

Perimeter:_____.

Area:_____.

3)

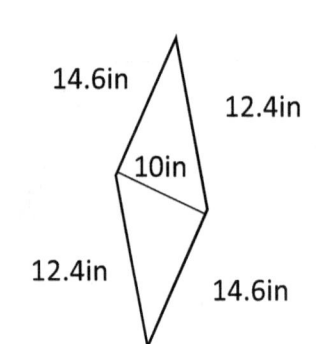

Perimeter:_____.

Area _____.

4)

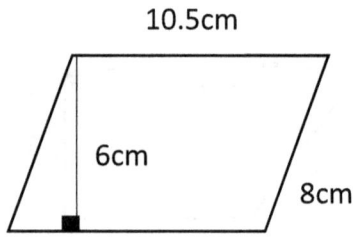

Perimeter:_____.

Area:_____.

5)

24.5m

18m

Perimeter:_____.

Area:_____.

6)

12 m

Perimeter:_____.

Area:_____.

Circumference and Area of Circle

Find the circumference and area of each ($\pi = 3.14$).

1)

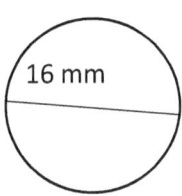

Circumference:

Area:

2)

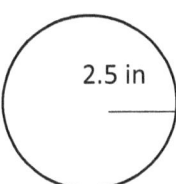

Circumference:⎯⎯⎯⎯.

Area:⎯⎯⎯.

3)

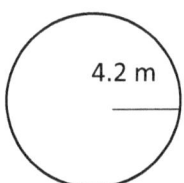

Circumference:⎯⎯⎯⎯.

Area ⎯⎯.

4)

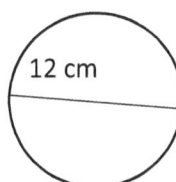

Circumference:⎯⎯⎯⎯⎯:

Area:⎯⎯⎯⎯.

5)

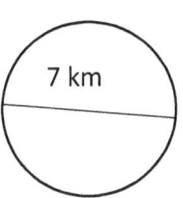

Circumference:⎯⎯⎯.

Area:⎯⎯⎯.

6)

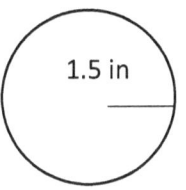

Circumference:⎯⎯⎯⎯⎯:

Area:⎯⎯⎯:

Perimeter of Polygon

Find the perimeter of each polygon.

1)

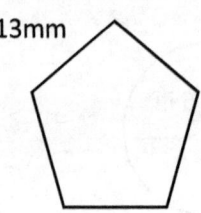

Perimeter: _____ .

2)

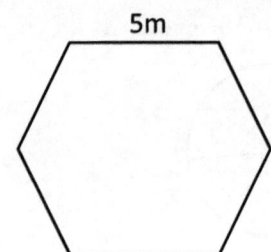

Perimeter: _____ .

3)

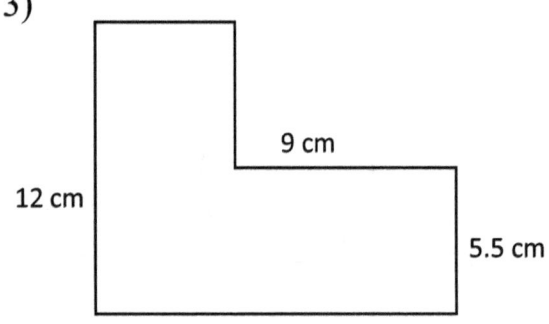

Perimeter: _____ .

4)

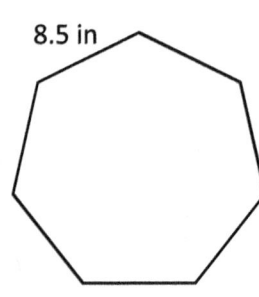

Perimeter: _____ .

5)

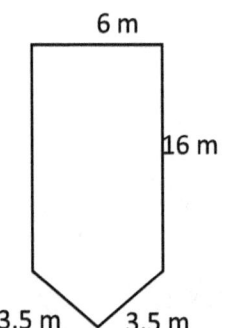

Perimeter: _____ .

6)

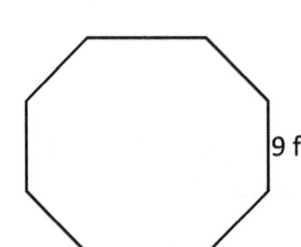

Perimeter: _____ .

Volume of Cubes

Find the volume of each cube.

1)

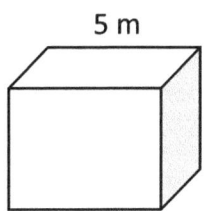

V: ....

2)

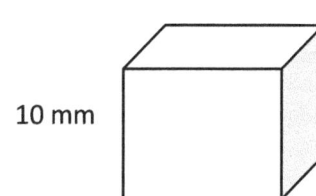

V: ....

3)

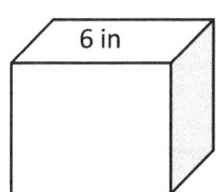

V: ....

4)

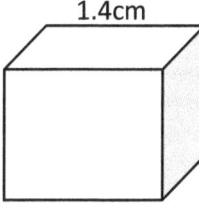

V: ....

5)

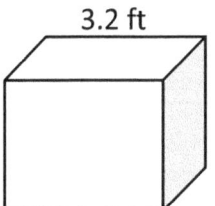

V: ....

6)

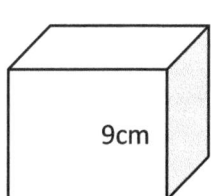

V: ....

SBAC Math Practice Grade 6

Volume of Rectangle Prism

Find the volume of each rectangle prism

1)

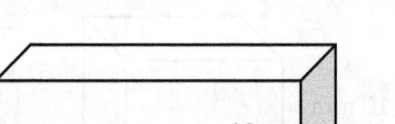

V:_____.

2)

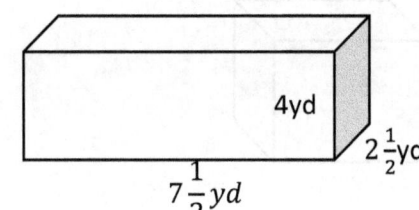

V:_____.

3)

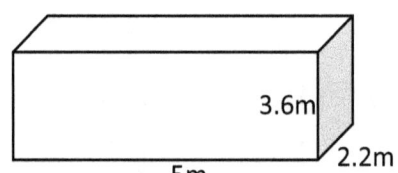

V:_____.

4)

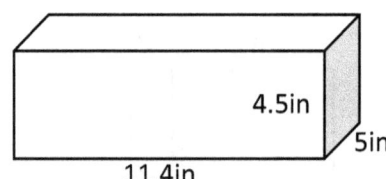

V:_____.

5)

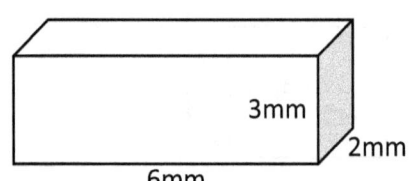

V:_____.

6)

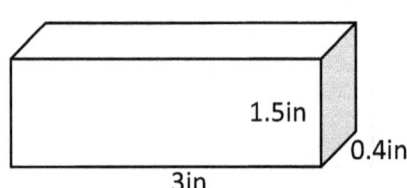

V:_____.

Volume of Cylinder

Find the volume of each cylinder.

1)
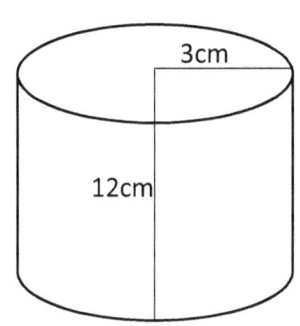

V:_____:

2)
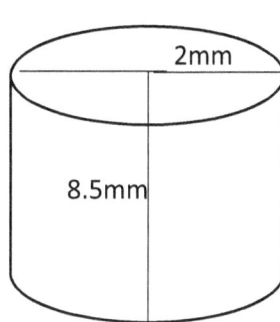

V:_____:

3)
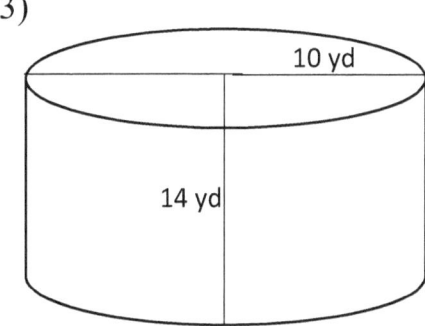

V:_____:

4)
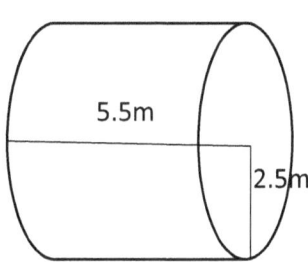

V:_____:

5)
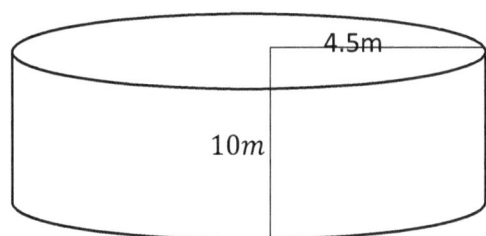

V:_____:

6)
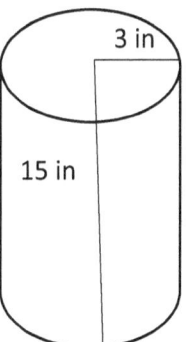

V:_____:

Surface Area Cubes

Find the surface area of each cube.

1)

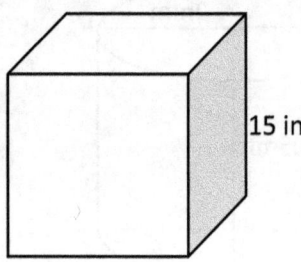
15 in

SA: _____ .

2)

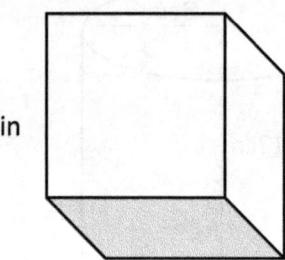
8 in

SA: _____ .

3)

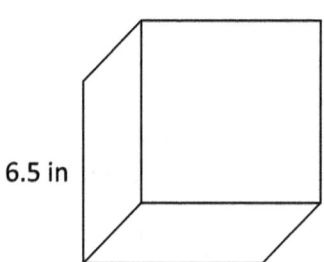
6.5 in

SA: _____ .

4)

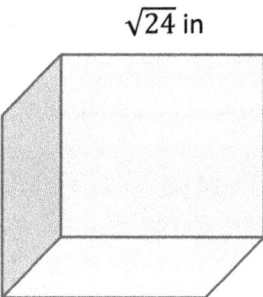
$\sqrt{24}$ in

SA: _____ .

5)

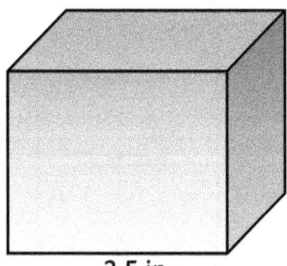
3.5 in

SA: _____ .

6)

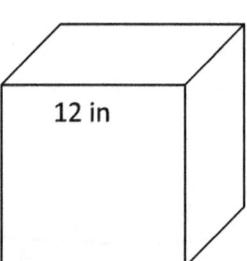
12 in

SA: _____ .

Surface Area Rectangle Prism

Find the surface area of each rectangular prism.

1)

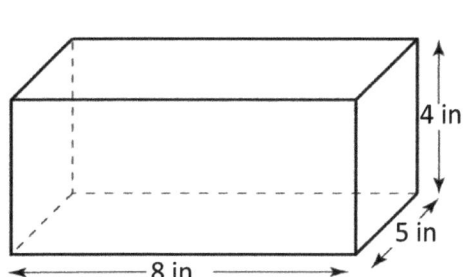

SA:_____.

2)

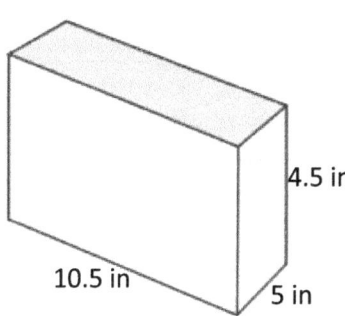

SA:_____.

3)

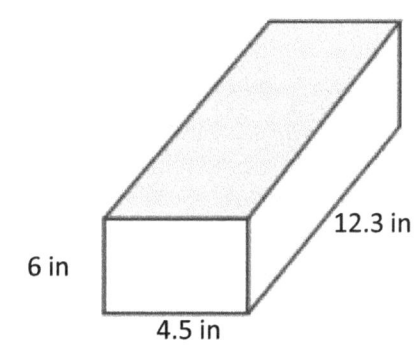

SA:_____.

4)

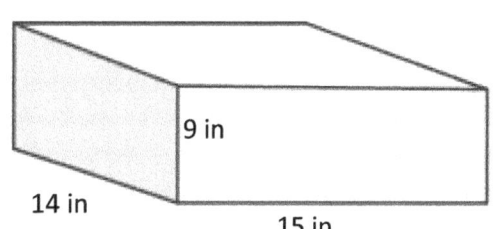

SA:_____.

5)

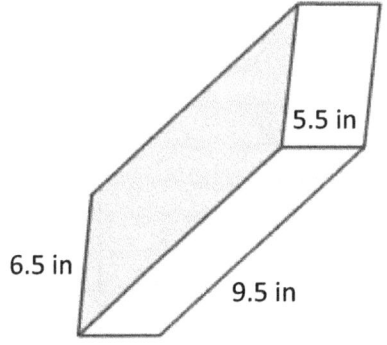

SA:_____.

6)

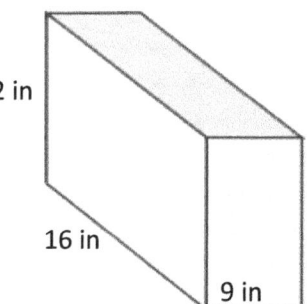

SA:_____.

SBAC Math Practice Grade 6

Surface Area Cylinder

Find the surface area of each cylinder.

1)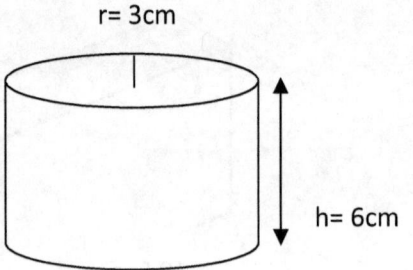

SA:_____:

2)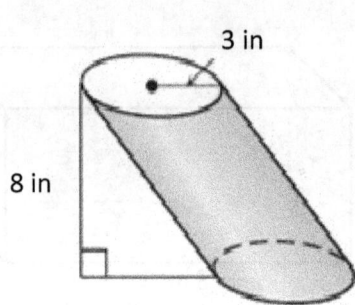

SA:_____:

3)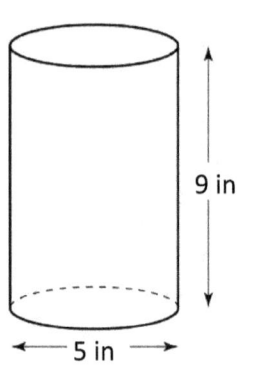

SA:_____:

4)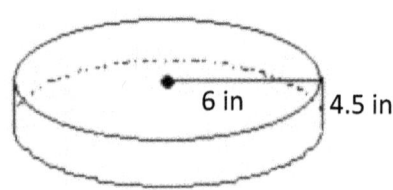

SA:_____:

5)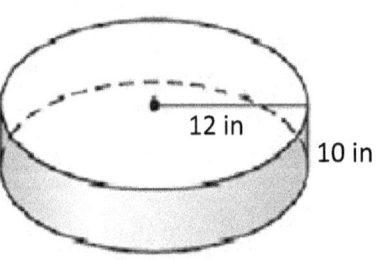

SA:_____:

6)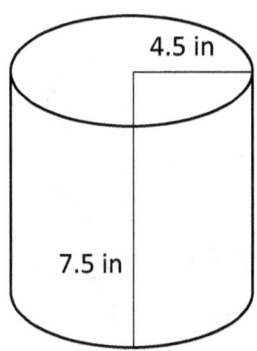

SA:_____:

Answer key Chapter 10

Area and Perimeter of Square

1. Perimeter: 20, Area: 25
2. Perimeter: $4\sqrt{6}$, Area: 6
3. Perimeter: 28, Area: 49
4. Perimeter: $4\sqrt{9}$, Area: 9
5. Perimeter: 48, Area: 144
6. Perimeter: $4\sqrt{50}$, Area: 50

Area and Perimeter of Rectangle

1- Perimeter: 18, Area: 18
2- Perimeter: 40, Area: 96
3- Perimeter: 50, Area: 150
4- Perimeter: 19, Area: 17.5
5- Perimeter: 4.23, Area: 1
6- Perimeter: 14, Area: 10

Area and Perimeter of Triangle

1- Perimeter: 3s, Area: $\frac{1}{2}sh$
2- Perimeter: 96, Area: 384
3- Perimeter: 33, Area: 45
4- Perimeter: 24, Area: 24
5- Perimeter: 24, Area: 24
6- Perimeter: 47.9, Area: 97.5

Area and Perimeter of Trapezoid

1- Perimeter: 32, Area: 49
2- Perimeter: 22, Area: 26
3- Perimeter: 44, Area: 93
4- Perimeter: 50, Area: 144
5- Perimeter: 59, Area: 245
6- Perimeter: 23, Area: 17

Area and Perimeter of Parallelogram

1- Perimeter: $26m$, Area: $20(m)^2$
2- Perimeter: $50m$, Area: $78(m)^2$
3- Perimeter: $54in$, Area: $146(in)^2$
4- Perimeter: $37cm$, Area: $63(cm)^2$
5- Perimeter: $85m$, Area: $441(m)^2$
6- Perimeter: $48m$, Area: $144(m)^2$

Circumference and Area of Circle

1) Circumference: 50.24 mm   Area: $200.96(mm)^2$
2) Circumference: 15.7in   Area: $(19.63in)^2$
3) Circumference: 26.38 m   Area: $55.39(m)^2$
4) Circumference: 37.68 cm   Area: 113.04
5) Circumference: 21.98 in   Area: $38.47(in)^2$
6) Circumference: 9.42 km   Area: $7.07(km)^2$

Perimeter of Polygon

1) 65 mm
2) 30 m
3) 57 cm
4) 59.5 in
5) 45 m
6) 72 ft

Volume of Cubes

1) $125m^3$
2) $1,000(mm)^3$
3) $216in^3$
4) $2.74(cm)^3$

5) $32.77 (ft)^3$ 6) $729 (cm)^3$

Volume of Rectangle Prism

1) $720 (cm)^3$ 3) $39.6 (m)^3$ 5) $36 (mm)^3$

2) $75 (yd)^3$ 4) $256.5 (in)^3$ 6) $1.8 (in)^3$

Volume of Cylinder

1) $339.12 (cm)^3$ 3) $1,099 (yd)^3$ 5) $635.85 (m)^3$

2) $26.69 (mm)^3$ 4) $107.94 (m)^3$ 6) $423.9 (in)^3$

Surface Area Cubes

1) $1,350 (in)^2$ 3) $253.5 (in)^2$ 5) $73.5 (in)^2$

2) $384 (in)^2$ 4) $144 (in)^2$ 6) $864 (in)^2$

Surface Area Rectangle Prism

1) $184 (in)^2$ 3) $312.3 (in)^2$ 5) $299.5 (in)^2$

2) $244.5 (in)^2$ 4) $942 (in)^2$ 6) $888 (in)^2$

Surface Area Cylinder

1) $169.56 (in)^2$ 3) $180.55 (in)^2$ 5) $1,657.92 (in)^2$

2) $207.24 (in)^2$ 4) $395.64 (in)^2$ 6) $339.12 (in)^2$

Chapter 11:
Statistics and probability

Mean, Median, Mode, and Range

Find the mean, median, mode(s), and range of the following data.

1) 26, 69, 30, 27, 19, 54, 27

Mean: __, Median: __, Mode: __, Range: __

2) 8, 12, 12, 15, 18, 20

Mean: __, Median: __, Mode: __, Range: __

3) 51, 32, 29, 33, 39, 17, 25, 29, 12

Mean: __, Median: __, Mode: __, Range: __

4) 10, 7, 3, 9, 2, 4

Mean: __, Median: __, Mode: __, Range: __

5) 20, 16, 10, 19, 13, 18, 12, 9, 9, 7

Mean: __, Median: __, Mode: __, Range: __

6) 9, 17, 18, 9, 6, 18, 8, 12

Mean: __, Median: __, Mode: __, Range: __

7) 49, 48, 86, 96, 34, 64, 48, 14, 32, 64

Mean: __, Median: __, Mode: __, Range: __

8) 45, 45, 47, 88, 89

Mean: __, Median: __, Mode: __, Range: __

9) 18, 18, 28, 36, 64

Mean: __, Median: __, Mode: __, Range: __

10) 10, 8, 2, 2, 5, 8, 1

Mean: __, Median: __, Mode: __, Range: __

11) 5, 9, 3, 5, 1, 7

Mean: __, Median: __, Mode: __, Range: __

12) 6, 7, 11, 11, 12, 12, 12

Mean: __, Median: __, Mode: __, Range: __

13) 8, 8, 0, 16, 0, 8, 16

Mean: __, Median: __, Mode: __, Range: __

14) 12, 18, 20, 7, 11, 10, 12, 16

Mean: __, Median: __, Mode: __, Range: __

15) 6, 12, 15, 15, 20

Mean: __, Median: __, Mode: __, Range: __

16) 9, 9, 12, 10, 12, 8, 17

Mean: __, Median: __, Mode: __, Range: __

17) 20, 8, 6, 9, 18, 19, 9, 6

Mean: __, Median: __, Mode: __, Range: __

18) 62, 16, 16, 28, 3, 2

Mean: __, Median: __, Mode: __, Range: __

19) 55, 22, 24, 55, 2, 4

Mean: __, Median: __, Mode: __, Range: __

20) 98, 64, 73, 86, 91, 98, 79

Mean: __, Median: __, Mode: __, Range: __

Box and Whisker Plot

1) Draw a box and whisker plot for the data set:

   24, 21, 22, 26, 24, 22, 26, 26, 30

2) The box-and-whisker plot below represents the math test scores of 20 students.

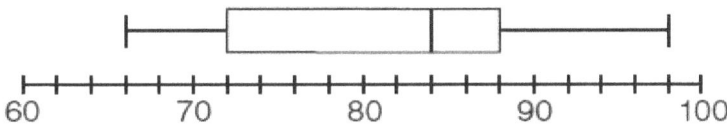

   A. What percentage of the test scores are less than 72?

   B. Which interval contains exactly 50% of the grades?

   C. What is the range of the data?

   D. What do the scores 66, 84, and 98 represent?

   E. What is the value of the lower and the upper quartile?

   F. What is the median score?

Bar Graph

Each student in class selected two games that they would like to play. Graph the given information as a bar graph and answer the questions below:

Game	Votes
Football	12
Volleyball	9
Basketball	15
Baseball	19
Tennis	15

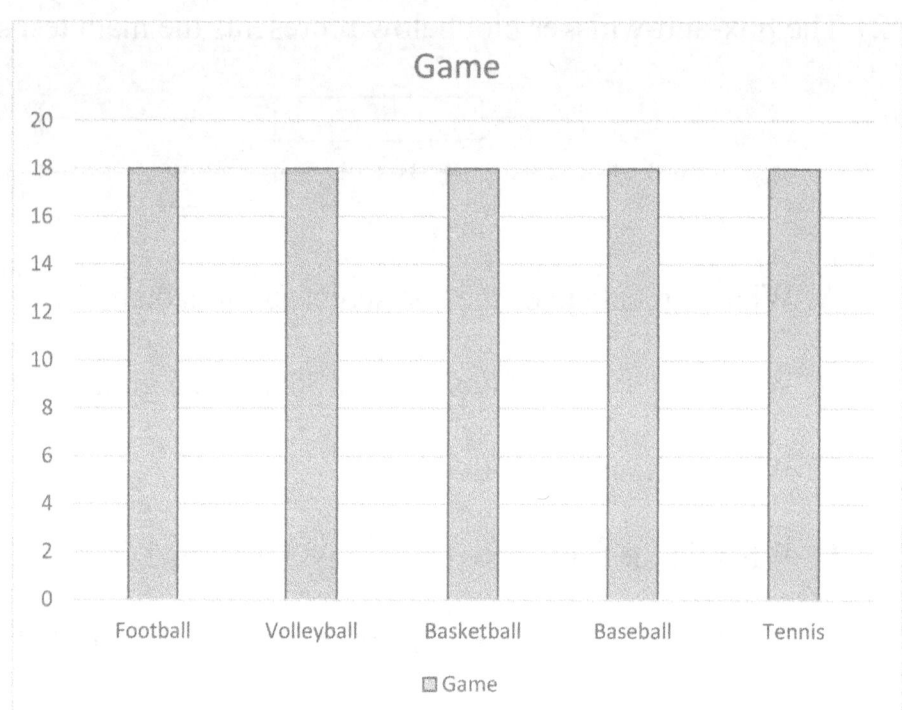

1) Which was the most popular game to play?

2) How many more student like Baseball than Football?

3) Which two game got the same number of votes?

4) How many Volleyball and Football did student vote in all?

5) Did more student like football or Tennis?

6) Which game did the fewest student like?

Histogram

Create a histogram for the set of data.

Math Test Score out of 100 points.

68	84	73	90	93	75	80	96	77	64
91	83	92	85	81	66	97	76	84	82
94	65	86	83	77	95	79	78	62	97

Frequency Table	
Interval	Number of Values
60–69	5
70–79	7
80–89	9
90–99	9

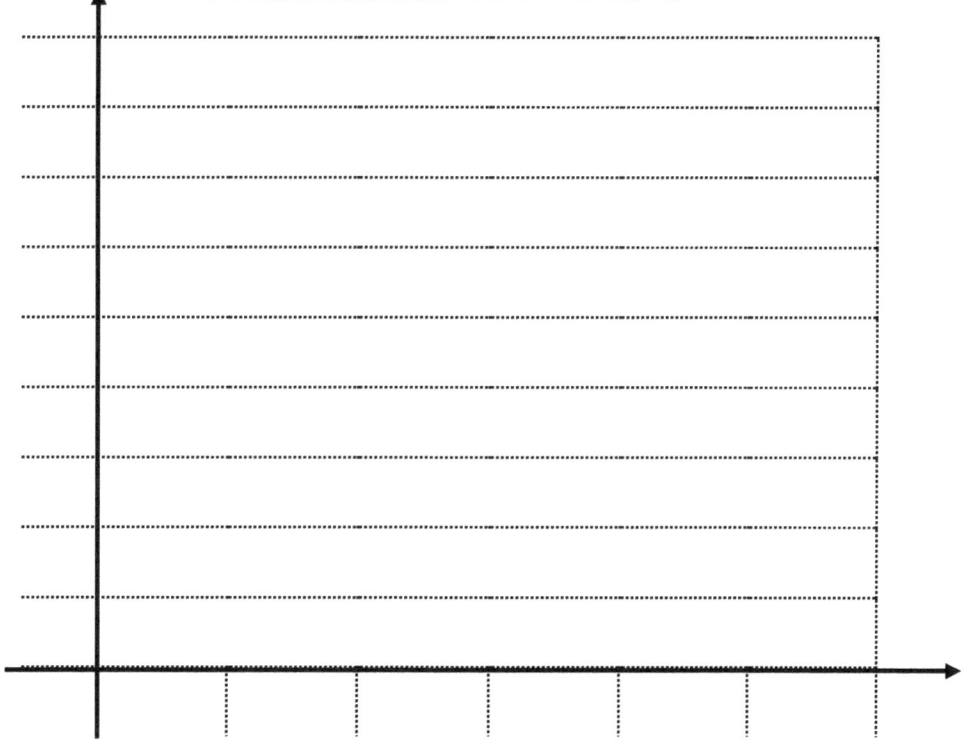

Dot plots

The ages of students in a Math class are given below.

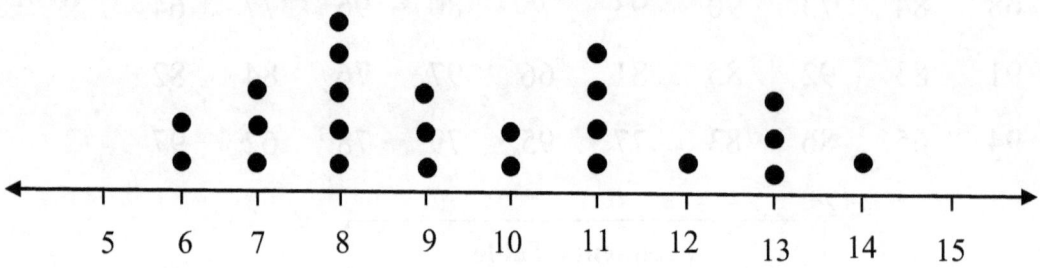

1) What is the total number of students in math class?

2) How many students are at least 11 years old?

3) Which age(s) has the most students?

4) Which age(s) has the fewest student?

5) Determine the median of the data.

6) Determine the range of the data.

7) Determine the mode of the data.

Stem-And-Leaf Plot

Make stem-and-leaf plots for the given data.

1) 22, 26, 28, 21, 42, 24, 48, 47, 29, 24, 19, 12, 45

2) 52, 54, 27, 31, 52, 24, 36, 58, 38, 34, 39, 32

Stem	leaf

3) 113, 106, 95, 95, 100, 115, 92, 114, 98, 112, 96, 107

Stem	leaf

4) 22, 15, 27, 21, 79, 24, 70, 77, 29, 24, 19, 12

Stem	leaf

5) 66, 69, 123, 67, 19, 126, 120

Stem	leaf

6) 112, 87, 96, 85, 110, 117, 92, 114, 88, 112, 98, 90

Stem	leaf

SBAC Math Practice Grade 6

Pie Graph

80 people were survey on their favorite ice cream. The pie graph is made according to their responses. Answer following questions based on the Pie graph.

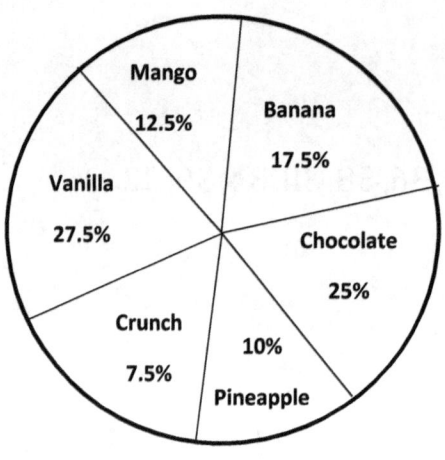

1) How many people like to eat Banana ice cream? _____

2) Approximately, which two ice creams did about half the people like the best? _____

3) How many people said either mango or crunch ice cream was their favorite? _____

4) How many people would like to have chocolate ice cream? _____

5) Which ice cream is the favorite choice of 22 people? _____

Probability

1) A jar contains 12 caramels, 7 mints and 16 dark chocolates. What is the probability of selecting a mint?

2) If you were to roll the dice one time what is the probability it will NOT land on a 2?

3) A die has sides are numbered 1 to 6. If the cube is thrown once, what is the probability of rolling a 6?

4) The sides of number cube have the numbers 3, 5, 7, 3, 5, and 7. If the cube is thrown once, what is the probability of rolling a 5?

5) Your friend asks you to think of a number from eight to twenty. What is the probability that his number will be 13?

6) A person has 5 coins in their pocket. A dime, 2 pennies, a quarter, and a nickel. If a person randomly picks one coin out of their pocket. What would the probability be that they get a penny?

7) What is the probability of drawing an odd numbered card from a standard deck of shuffled cards?

8) 24 students apply to go on a school trip. Three students are selected at random. what is the probability of selecting 3 students?

SBAC Math Practice Grade 6

Answer key Chapter 11

Mean, Median, Mode, and Range of the Given Data

1) mean: 36, median: 27, mode: 27, range: 50
2) mean: 14.17, median: 13.5, mode: 12, range: 12
3) mean: 29.7, median: 29, mode: 29, range: 39
4) mean: 5.83, median: 5.5, mode No mode. range: 8
5) mean: 13.3, median: 12.5, mode: 9, range: 13
6) mean: 12.125, median: 10.5, mode: 9,18, range: 12
7) mean: 53.5, median: 48.5, mode: 48 and 64, range: 82
8) mean: 62.8, median: 47, mode: 45, range: 44
9) mean: 32.8, median: 28, mode: 18, range: 46
10) mean: 5.1, median: 5, mode: 2,8, range: 9
11) mean: 5, median: 5, mode: 5, range: 8
12) mean: 10.14, median: 11, mode: 12, range: 6
13) mean: 8, median: 8, mode: 8, range: 16
14) mean: 13.25, median: 12, mode: 12, range: 13
15) mean: 13.6, median: 15, mode: 15, range: 14
16) mean: 11, median: 10, mode: 9,12, range: 9
17) mean: 11.88, median: 9, mode: 6,9, range: 14
18) mean: 21.17, median: 16, mode: 16, range: 60
19) mean: 27, median: 23, mode: 55, range: 53
20) mean: 84.14, median: 86, mode: 98, range: 34

Box and Whisker Plot

1)

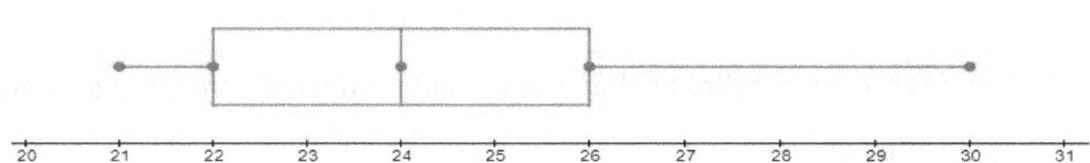

2)
- A. 25%
- B. 72-84
- C. 32
- D. Minimum, Median, and Maximum

E. Lower (Q_1) is 72 and upper (Q_3) is 88   F. 84

Bar Graph

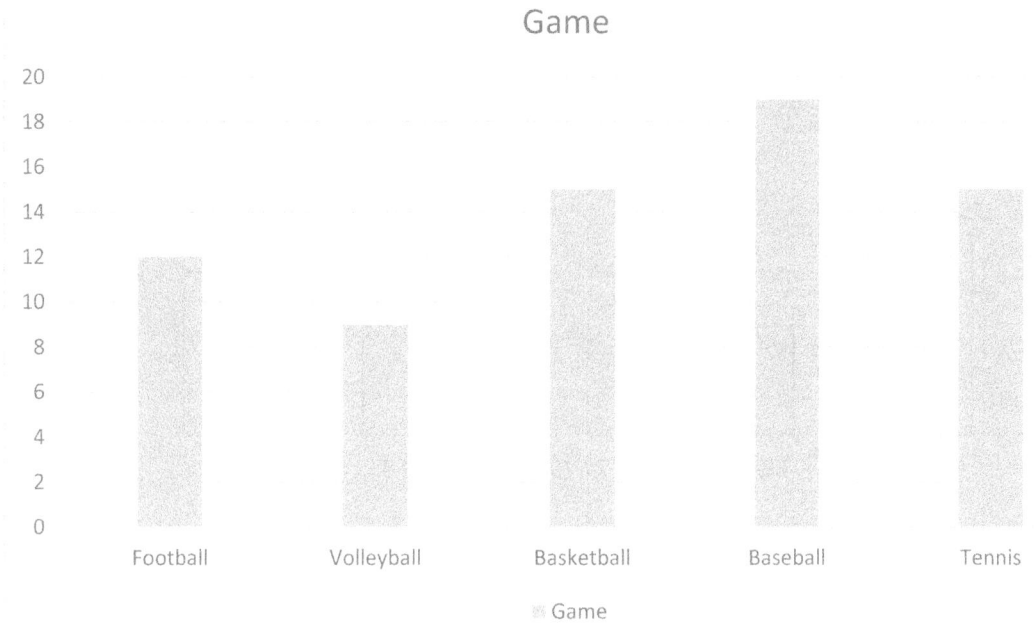

1) Baseball
2) 7 students
3) Basketball and Tennis
4) 21
5) Tennis
6) Volleyball

Histogram

Frequency Table	
Interval	Number of Values
62-67	4
68-73	2
74-79	6
80-85	8
86-91	3
92-97	7

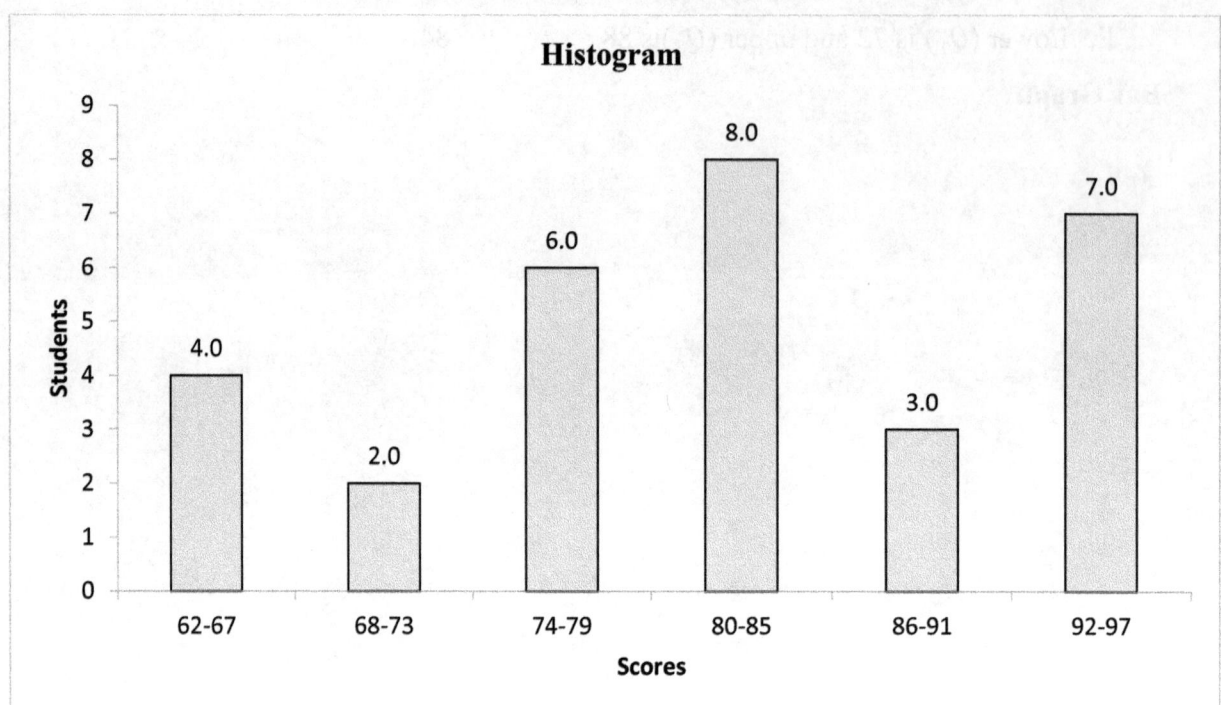

Dot plots

1) 24
2) 9
3) 8
4) 12 and 14
5) 3
6) 8
7) 3

Stem–And–Leaf Plot

1)

Stem	leaf
1	2 9
2	1 2 4 4 6 8 9
4	2 7 8 5

2)

Stem	leaf
2	4 7
3	1 2 4 6 8 9
5	2 2 4 8

3)

Stem	leaf
9	2 5 5 6 8
10	0 6 7
11	2 3 4 5

4)

Stem	leaf
1	2 9 5
2	1 2 4 4 7 9
7	0 7 9

5)

Stem	leaf
1	9
6	6 7 9
12	0 3 6

6)

Stem	leaf
8	5 7 8
9	0 2 6 8
11	0 2 2 4 7

Pie Graph

1) 14
2) Vanilla and chocolate
3) 16
4) 20
5) Vanilla

Probability

1) $\frac{1}{5}$
2) $\frac{5}{6}$
3) $\frac{1}{6}$
4) $\frac{1}{3}$
5) $\frac{1}{12}$
6) $\frac{2}{5}$
7) $\frac{4}{13}$
8) $\frac{1}{8}$

SBAC Test Review

SBAC GRADE 6 MAHEMATICS REFRENCE MATERIALS

Conversions:

LENGTH

Customary	Metric
1 mile (mi) = 1,760 yards (yd)	1 kilometer (km) = 1,000 meters (m)
1 yard (yd) = 3 feet (ft)	1 meter (m) = 100 centimeters (cm)
1 foot (ft) = 12 inches (in.)	1 centimeter (cm) = 10 millimeters (mm)

VOLUME AND CAPACITY

Customary	Metric
1 gallon (gal) = 4 quarts (qt)	1 liter (L) = 1,000 milliliters (mL)
1 quart (qt) = 2 pints (pt.)	
1 pint (pt.) = 2 cups (c)	
1 cup (c) = 8 fluid ounces (Fl oz)	

WEIGHT AND MASS

Customary	Metric
1 ton (T) = 2,000 pounds (lb.)	1 kilogram (kg) = 1,000 grams (g)
1 pound (lb.) = 16 ounces (oz)	1 gram (g) = 1,000 milligrams (mg)

Formulas:

Area

Triangle	$A = \frac{1}{2}bh$
Rectangle or Parallelogram	$A = bh$
Trapezoid	$A = \frac{1}{2}h(b_1 + b_2)$

Volume

Rectangular Prism	$V = Bh$

Smarter Balanced Assessment Consortium

SBAC Practice Test 1

Mathematics

GRADE 6

- ❖ 30 Questions
- ❖ There is no time limit for this practice test.
- ❖ Basic Calculators are permitted for this practice test

Administered *Month Year*

SBAC Math Practice Grade 6

1) Which of the following fractions is closest to zero?

   A. $-\frac{1}{5}$

   B. $-\frac{3}{8}$

   C. $-\frac{3}{5}$

   D. $-\frac{1}{4}$

2) Enter the unknown value that makes this statement true:

   $$30\% \text{ of } \square \text{ is } 48$$

   A. 106

   B. 160

   C. 120

   D. 240

3) Ms. Carson buys groceries for a total of $71.25. She now has 20.84 left. Which equation could be used to find out how much money Ms. Carson had before she bought the groceries?

   A. $x + \$20.84 = \$71.25$

   B. $x + \$71.25 = \$20.84$

   C. $x - \$71.25 = \$20.84$

   D. $\$20.84x = \$71.25$

4) $(-9) - (-6) =$

    A. 3

    B. 15

    C. −15

    D. −3

5) Look at the equation. $\frac{4}{9} \times \frac{\square}{\square} = n$

    Emma claims that for any fraction multiplied by $\frac{4}{9}$, n will be less than $\frac{4}{9}$. Which number convince Emma that this statement is **NOT** true.

    A. $\frac{1}{4}$

    B. $\frac{9}{8}$

    C. $\frac{3}{4}$

    D. $\frac{5}{8}$

6) Carl types 168 words in 4 minutes. Enter the number of words carl types in 6 minutes at this rate.

    A. 525

    B. 200

    C. 252

    D. 260

7) After 9 new orders placed in a restaurant, the restaurant had 54 orders. Which equation can be used to find n, number of orders in the restaurant before the 9 new orders placed?

A. $9n = 54$

B. $n + 9 = 54$

C. $\frac{n}{9} = 54$

D. $n - 9 = 54$

8) Which expression shows a prime factorization?

A. $1 \cdot 7 \cdot 11$

B. $2 \cdot 5 \cdot 18$

C. $3.5 \cdot 11 \cdot 13$

D. $2 \cdot 2 \cdot 7 \cdot 7 \cdot 11$

9) The height of four kids are shown in inches.

$$7.07 \quad 7\frac{2}{5} \quad 7.15 \quad 7\frac{3}{8}$$

Which list shows these heights in order from greatest to least?

A. $7\frac{2}{5}$; $7\frac{3}{8}$; 7.15; 7.07

B. 7.15; $7\frac{2}{5}$; 7.07; $7\frac{3}{8}$

C. 7.07; 7.15; $7\frac{3}{8}$; $7\frac{2}{5}$

D. $7\frac{2}{5}$; 7.15; 7.07; $7\frac{3}{8}$

10) Alex has $\frac{5}{6}$ pound of bird food. He puts an equal portion into 3 bird feeders. How much food, in pounds, does he put into each bird feeder?

A. $\frac{6}{7}$

B. $1\frac{1}{8}$

C. $\frac{5}{18}$

D. $2\frac{3}{8}$

11) Which number line shows the correct locations of all given values?

A.

B.

C.

D.

12) Select the expression that is equivalent to $15(t + 5)$.

A. $(15 + 5t) + (15 + 3t)$

B. $5(3t + 15)$

C. $5t + 5 + 5t$

D. $(15 \times t) \times (15 \times 5)$

13) Sea level is 0 feet in elevation. The elevation of land represents its height above or below sea level. This table shows the lowest elevation in some states. Determine whether each statement about the lowest elevation is correct.

State	Lowest Elevation (ft)
Kentucky	119
Louisiana	−41
California	−184
Arkansas	66

A. The elevation at the lowest point in California is higher than the lowest point in Kentucky.

B. The elevation at the lowest point in Louisiana is higher than the lowest point in Arkansas.

C. The elevation at the lowest point in Louisiana is farther from 0 than the elevation at the lowest point in Arkansas.

D. The elevation at the lowest point in California is farther from 0 than the elevation at the lowest point in Kentucky.

14) If O is a point on the AB line, what is the value in degrees of the DOC angle?

A. 70°

B. 60°

C. 50°

D. 130°

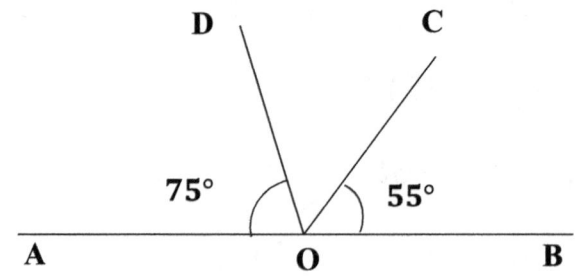

15) Bernard needs 3.1 meters of wire for one project and 0.8 meter of wire for another project. If Bernard has 3.4 meters of wire, how much wire he will have left over or how much more he needs?

A. 0.4

B. 0.5

C. -0.4

D. -0.5

16) Mr. Hendrik went out to lunch and the cost of the food was $64. If the tip given was 25%, what was the total cost?

A. $25

B. $80

C. $90

D. $125

17) The coordinates of parallelogram are given below. Find the x-coordinates of vertex D(d, 3).

A. $d = 5$

B. $d = -2$

C. $d = 4$

D. $d = 1$

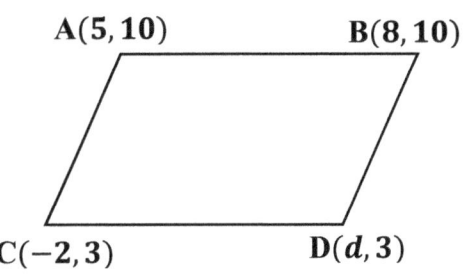

18) A recipe requires $\frac{5}{7}$ cup of nuts for 1 cake. What is the maximum number of cakes that can made using $4\frac{2}{7}$ cups of nuts?

A. 6

B. 5

C. 12

D. 4

19) Select the value that completes this expression for converting 19 meters to millimeters.

$$\left(\frac{19\ m}{1}\right)\left(\boxed{}\right)\left(\frac{10\ mm}{1\ cm}\right)$$

A. $\left(\frac{19m}{1,000\ mm}\right)$

B. $\left(\frac{100\ cm}{1\ m}\right)$

C. $\left(\frac{1,000\ mm}{19\ m}\right)$

D. $\left(\frac{1\ cm}{100\ mm}\right)$

20) Which arithmetic sequence is represented by the expression $3n - 2$, where m represents the position of a term in the sequence?

A. 8, 11, 14, 17, 20, …

B. 15, 18, 21, 24, 27, …

C. 7, 11, 15, 19, 23, …

D. 7, 10, 13, 16, 19, …

21) Which diagram best represents the relationship among integers, rational numbers, whole numbers and −119?

A.
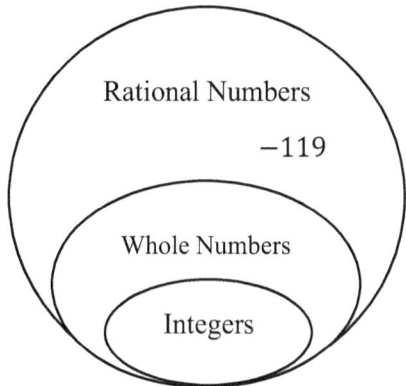

C.
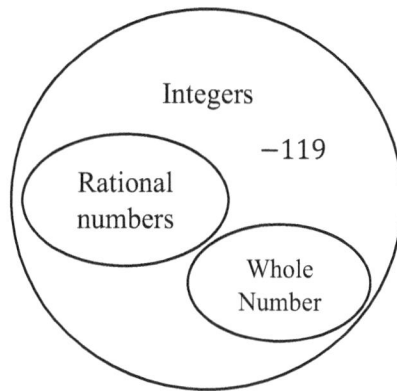

B.
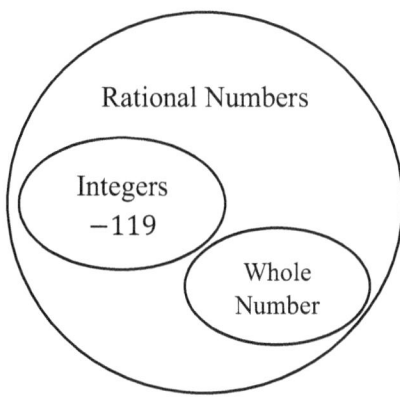

D.
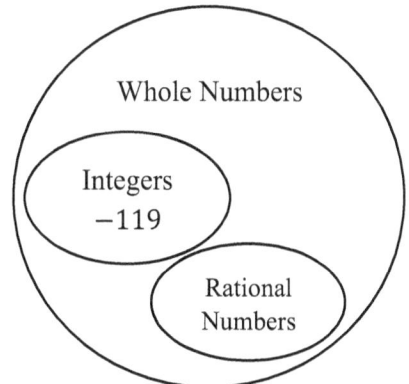

22) What is the equivalent temperature of $167°F$ in Celsius?

$$C = \frac{5}{9}(F - 32)$$

A. $75°C$

B. $12°C$

C. $400°C$

D. $157°C$

23) The dot plots show how many minutes per day do 6th grade study math after school at two different schools on one day.

Number of minuets study in school 1

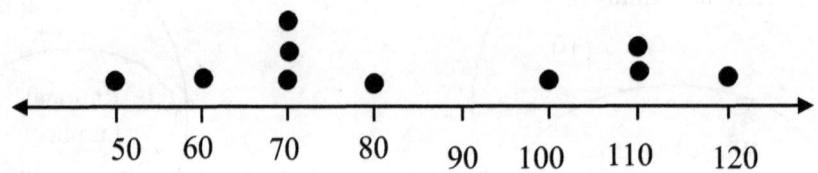

Number of minuets study in school 2

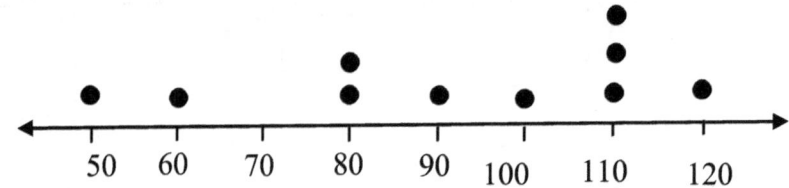

Which statement is supported by the information in the dot plots?

A. The mode of the data for School 2 is greater than the mode of the data for School 1.

B. The mean of the data for School 1 is greater than the mean of the data for School 2.

C. The median of the data for School 2 is smaller than the median of the data for School 1.

D. The median and mean of the data for two schools are equal.

24) Compare the numbers $|-57|\,\square\,-57$

A. $\geq$

B. $<$

C. $\leq$

D. $>$

25) The average of four numbers is 32. If a fifth number 47 is added, then, what is the new average?

A. 40

B. 30

C. 35

D. 20

26) One ordered pair is shown on a coordinate grid. What is the correct location of $B(+d, -f)$ $d, f > 0$ on the coordinate grid?

A. First quadrant (Q_I)

B. Second quadrant (Q_{II})

C. Third quadrant (Q_{III})

D. Fourth quadrant (Q_{IV})

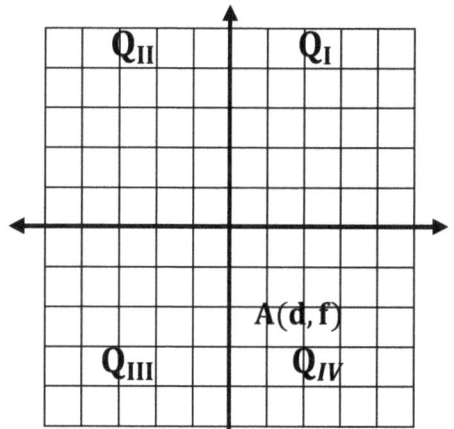

27) Joshua started to fill the box with some unit cubes. Find the total number of unit cubes needed to completely fill the box. Each unit cube is 1 cubic foot.

A. 63

B. 135

C. 495

D. 945

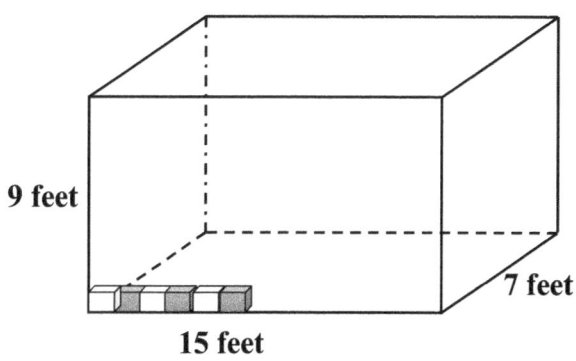

28) 5.6% written as a decimal is:

A. 0.56

B. 0.056

C. 0.0056

D. 5.6

29) The table contain x and y values in equivalent ratios. Fill in the missing value in the table.

A. 44

B. 42

C. 54

D. 46

x	y
6	36
7	
9	54
13	78

30) What percentage of the area of the square below is shadowed? All squares are equilateral.

A. 28%

B. 7%

C. 38%

D. 72%

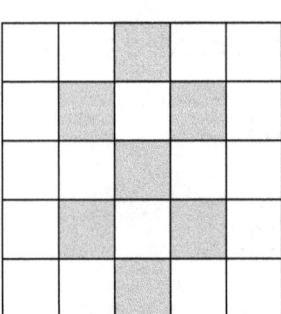

Smarter Balanced Assessment Consortium

SBAC Practice Test 2

Mathematics

GRADE 6

- ❖ 30 Questions
- ❖ There is no time limit for this practice test.
- ❖ Basic Calculators are permitted for this practice test

Administered *Month Year*

SBAC Math Practice Grade 6

1) Which of the descriptions below best describes the location of the point $(-2, 0)$?

   A. Quadrant II

   B. It is on the *x*-axis

   C. It is on the *y*-axis

   D. The coordinate planes

2) $0.25 \, km + 3,260 \, m + 8,000 \, cm = ?$

   A. $30.59 \, km$

   B. $0.33 \, km$

   C. $3.51 \, km$

   D. $3.59 \, km$

3) Which shows the expression rewritten in exponent form? $5 \times 5 \times 5 \times 5 \times 2 \times 2$

   A. $4^5 + 2^2$

   B. $5^4 + 2^2$

   C. $4^5 \times 2^2$

   D. $5^4 \times 2^2$

4) An isosceles trapezoid with sides 14, 7, 7, and 20 has a height of 16, what is the area?

   A. 272

   B. 227

   C. 98

   D. 172

5) Four points are labeled on the number line. Which point best represents $-\frac{1}{4}$

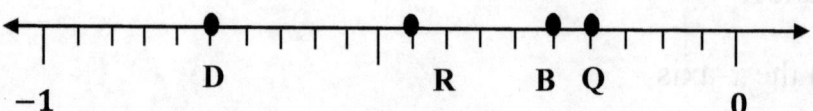

A. Point B

B. Point R

C. Point D

D. Point Q

6) The first number in a pattern is 23. Each following number is found by subtracting 7 from the previous number. What is the sixth number in the pattern?

   A. −37

   B. −13

   C. −5

   D. −12

7) Use the set of data below. What is the median of the list of numbers?

22, 14, 21, 15, 14, 17

   A. 16

   B. 14

   C. 18

   D. 21

8) A bird flies at a constant speed for 33 meters. If it takes 5 seconds to fly the first 7 meters, which is the equation that can be used to find the time t it takes the bird to fly the 33 yards?

A. $\dfrac{7}{5} = \dfrac{t}{33}$

B. $\dfrac{7}{5} = \dfrac{5}{t}$

C. $\dfrac{7}{33} = \dfrac{t}{5}$

D. $\dfrac{7}{33} = \dfrac{5}{t}$

9) The angle measures of a triangle GBD are shown in the diagram. What is the value of ∠B?

A. 34°

B. 84°

C. 89°

D. 146°

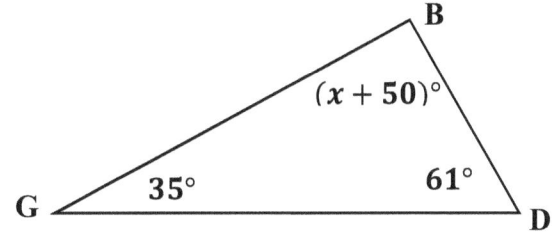

10) The three side lengths of an isosceles triangle are d, c and d. What equation can be used to find p, the perimeter of the triangle?

A. $p = d + 2c$

B. $p = \dfrac{1}{2}d + c$

C. $p = 2d + c$

D. $p = d + \dfrac{1}{2}c$

11) Which inequality is true if $p = 2.3$?

   A. $4p < 11.5$

   B. $15.9 \leq 5.2p$

   C. $7p > 18.6$

   D. $7.82 \geq 3.4p$

12) Look at the equation. $\frac{3}{8} \div \frac{\square}{\square} = n$

   Carl claims that for $\frac{3}{8}$ divided by any fraction, n will be less than $\frac{3}{8}$. Which number convince Carl that this statement is true.

   A. $\frac{3}{5}$

   B. $\frac{3}{4}$

   C. $\frac{1}{4}$

   D. 3

13) What is the area of parallelogram?

   A. $48\ cm^2$

   B. $96\ cm^2$

   C. $72\ cm^2$

   D. $36\ cm^2$

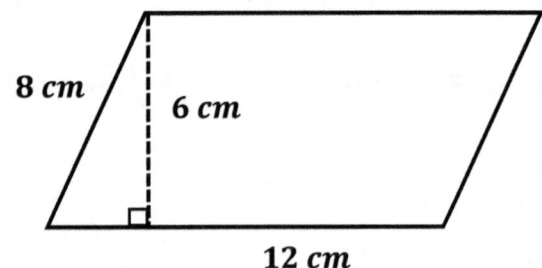

14) Which of the following is the opposite of the numbers $-N$, and 0?

   A. $N, -1$

   B. $\frac{1}{N}, 0$

   C. $\frac{1}{N}, -1$

   D. $N, 0$

15) Taylor pays $7.50 for 0.6 yard of fabric. What is the cost per yard?

   A. $2.15

   B. $15.02

   C. $12.50

   D. $102.00

16) $4(2x - 3) = ?$

   A. $4x - 3$

   B. $8x - 12$

   C. $4x - 12$

   D. $8x - 3$

17) Which expression represents the area?

   A. $bd + \frac{1}{2}a(c + d)$

   B. $bd + ad$

   C. $bd + \frac{1}{2}ac$

   D. $bd + ac$

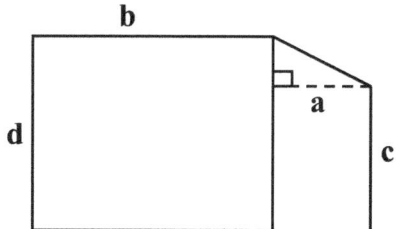

18) A computer company sells 9 notebook computers for every 5 desktop computers it sells. If the company sells 30 desktop computers in a day, how many notebook computers will it sell the same day?

A. 36

B. 45

C. 50

D. 54

19) The circle in the figure below has the radius equal to 9 cm. What is the ratio between the area and the circumference of the circle?

A. 4.5

B. 4

C. 3.50

D. 3

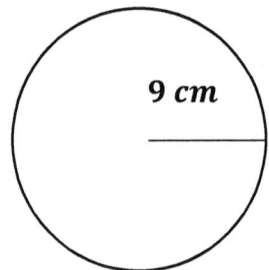

20) Mia had x cards and she gave $\frac{1}{5}$ to Sofia and Sofia gave $\frac{1}{6}$ of her card to Riley. Which algebraic expression that indicates the number of cards Sofia has?

A. $\frac{1}{3}x$

B. $\frac{1}{4}x$

C. $\frac{1}{6}x$

D. $\frac{1}{5}x$

21) The line in the figure below intersects the y-axis in the point with the coordinates:

A. $x = 0$ and $y = 8$

B. $x = 6$ and $y = 0$

C. $x = 0$ and $y = 6$

D. $x = 6$ and $y = 0$

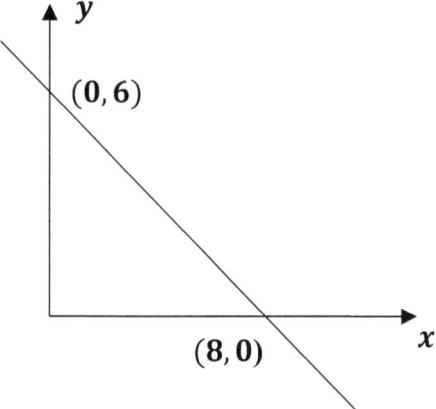

22) Use the chart below to answer the question.

outcome	red	blue	white	green
frequency	||||	|||| ||	||||	||

Nan randomly takes a marble from a bag, records the color in the chart, and then returns the marble to the bag. She does this 18 times and records her results in the chart each time. The chart shows her results. What is the experimental probability of drawing a white marble?

A. $\frac{1}{9}$

B. $\frac{4}{9}$

C. $\frac{2}{9}$

D. $\frac{5}{18}$

23) Use the coordinate grid below to answer the question. Which point best represents the coordinates (−3,4)?

A. L

B. N

C. M

D. P

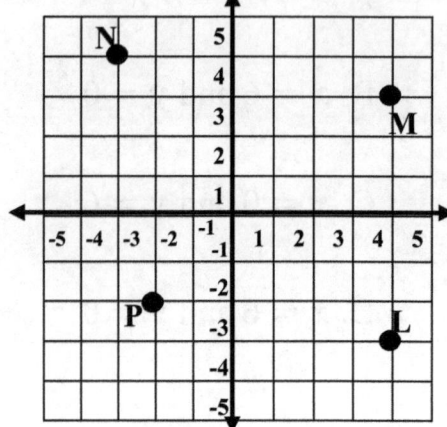

24) Everett has made $72.6 by selling lemonade. He charged $1.65 for a cup of lemonade. How many cups of lemonade did they sell?

A. 125

B. 64

C. 42

D. 44

25) Triangle MNP is similar to triangle BCD. What is the length of CD, if MN = 10 cm, NP = 8 cm and BD = 2.4 cm?

A. 3.00 cm

B. 3.5 cm

C. 3.75 cm

D. 4.5 cm

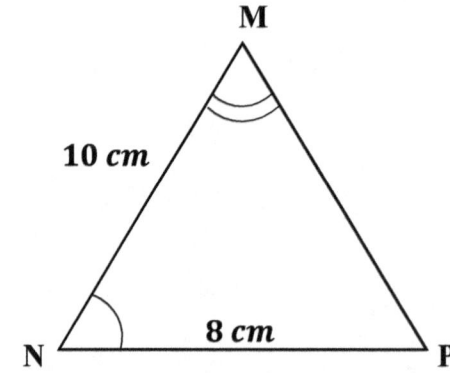
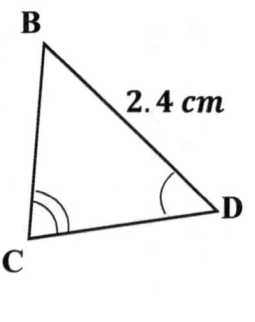

26) The following histogram summarizes the recorded number of books that a group of 240 students bought in a certain time period. Each interval contains possible values at the left endpoint up to but not including the right endpoint. What is the total number of students represented in the histogram who bought 7 or more books?

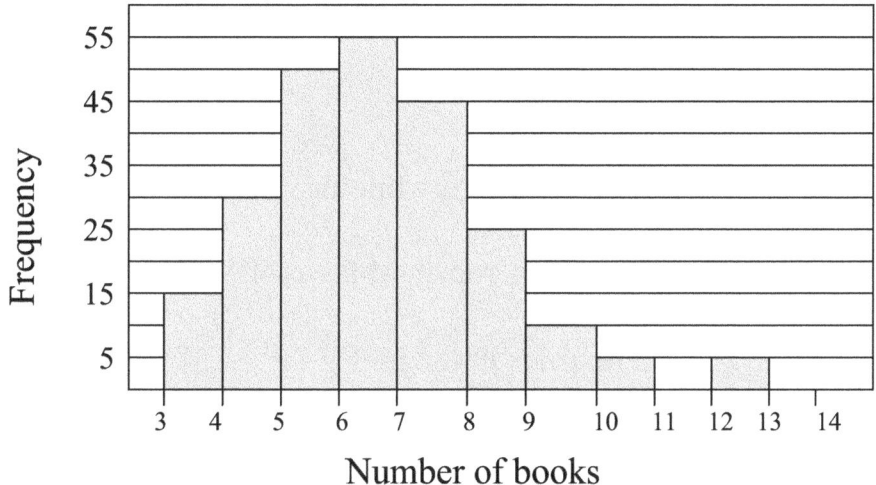

A. 85

B. 100

C. 90

D. 45

27) Use estimation to answer the question. Which sum is greater than 1?

A. $\frac{17}{28} + \frac{19}{35}$

B. $\frac{11}{28} + \frac{13}{35}$

C. $\frac{19}{45} + \frac{14}{39}$

D. $\frac{13}{45} + \frac{7}{39}$

28) What is the vertical line in the box of the box and whisker plot represented?

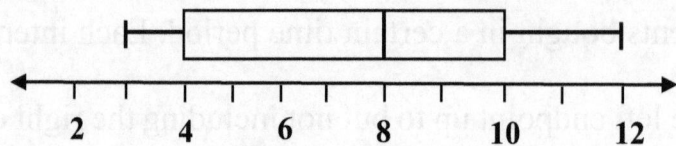

A. Range

B. Mean

C. Median

D. Interquartile Range

29) A jar contains some marbles that are white or red. The ratio of the number of white marbles to the number of red marbles is 4:5. What might be the total amount of white and red marbles in the jar?

A. 12 white marbles and 20 red marbles

B. 16 white marbles and 20 red marbles

C. 16 white marbles and 30 red marbles

D. 32 white marbles and 45 red marbles

30) Which ordered pair could belong to the ordered set A?

$$A = \{(6, 9.6), (7, 11.2), (8, 12.8)\}$$

A. (5, 8.6)

B. (5, 8)

C. (5, 8.4)

D. (5, 9)

Answers and Explanations

Answer Key

Now, it's time to review your results to see where you went wrong and what areas you need to improve!

SBAC Math Practice Tests

Practice Test 1

#	Ans	#	Ans
1	A	16	B
2	B	17	D
3	C	18	A
4	D	19	B
5	B	20	D
6	C	21	B
7	B	22	A
8	D	23	A
9	A	24	D
10	C	25	C
11	C	26	D
12	B	27	D
13	D	28	B
14	C	29	B
15	D	30	A

Practice Test 2

#	Ans	#	Ans
1	B	16	B
2	D	17	A
3	D	18	D
4	A	19	A
5	A	20	C
6	D	21	C
7	A	22	C
8	D	23	B
9	B	24	D
10	C	25	A
11	D	26	C
12	D	27	A
13	C	28	C
14	D	29	B
15	C	30	B

SBAC Math Practice Grade 6

SBAC Practice Test 1
Answers and Explanations

1) Answer: A

Change fraction to decimal:

$-\frac{1}{5} = -0.20$, $\quad -\frac{3}{8} = -0.375$, $\quad -\frac{3}{5} = -0.6$, $\quad -\frac{1}{4} = -0.25$

$-\frac{1}{5}$ is greatest, then close to 0. Or you just look at a dame number line.

2) Answer: B

$30\% \times \square = 48$ then, $0.30 \times \square = 48 \to \square = \frac{48}{0.30} = \frac{480}{3} = 160$

3) Answer: C

Let x be the total money she had, $x - 71.25 = 20.84$

4) Answer: D

Subtracting a number is the same as adding the opposite of the number.

$(-9) - (-6) = (-9) + 6 = -3$

5) Answer: B

Let's check the options provided:

A. $\frac{4}{9} \times \frac{1}{4} = \frac{4}{36} = \frac{1}{9} \to \frac{1}{9} < \frac{4}{9}$

B. $\frac{4}{9} \times \frac{9}{8} = \frac{36}{72} = \frac{1}{2} \to \frac{1}{2} \not< \frac{4}{9}$

C. $\frac{4}{9} \times \frac{3}{4} = \frac{12}{36} = \frac{1}{3} \to \frac{1}{3} < \frac{4}{9}$

D. $\frac{4}{9} \times \frac{5}{8} = \frac{20}{72} \to \frac{5}{18} < \frac{4}{9}$

6) Answer: C

If: 168 words = 4 minutes, then 1 minute = $\frac{168}{4} = 42$ words

6 minutes is: $6 \times 42 = 252$

7) Answer: B

You should have determined that "new" and "placed" means orders were added to the original numbers of orders and that the total of the original (n) and new orders (9) is 54, resulting in $n + 9 = 54$

8) Answer: D

Prime numbers are the numbers that have **only** two factors, 1 and itself. The number 1 is neither prime nor composite.

In this factorization, all the factors are prime numbers.

9) Answer: A

To determine the list that shows the heights in order from greatest (largest) to least (smallest), you should have changed the values in the list to the same form of number, resulting in a list of either all fractions or all decimals.

$7\frac{2}{5} = 7 + \frac{2}{5} = 7 + 0.4 = 7.4$, and $7\frac{3}{8} = 7 + \frac{3}{8} = 7 + 0.375 = 7.375$

Then order from greatest to least: $7.4, 7.375, 7.15, 7.07$

10) Answer: C

$\frac{5}{6} \div 3 = \frac{5}{6} \times \frac{1}{3} = \frac{5}{18}$

11) Answer: C

As you move to the right the numbers are positive and increase. As you go to the left, the numbers also increase and get more and more negative.

12) Answer: B

Use distribution property, then factor:

$15(t + 5) = (15 \times t) + (15 \times 5) = 5(3t) + 5(15) = 5(3t + 15)$

13) Answer: D

Compare each option:

A. $-184 > 119$ →Not correct

B. $-41 > 66$ → Not Correct

C. $|-41| > |66|$ →Not correct

D. $|-184| > |119|$ →Bingo!

14) Answer: C

$\angle AOB = 180°$(straight angle), and $\angle AOB = \angle AOD + \angle DOC + \angle COB$, then,

$180° = 75° + \angle DOC + 55° \rightarrow 180° = \angle DOC + 130°$

→ ∠DOC = 180° − 130° = 50°

15) Answer: D

$3.1 + 0.8 = 3.9; 3.4 − 3.9 = −0.5$

16) Answer: B

Use percent formula: part = $\frac{percent}{100}$ × whole

Tip = $\frac{25}{100} \times \$64 = \$16$. Then, the total payment is: $\$64 + \$16 = \$80$

17) Answer: D

The length is the horizontal distance between A (5, 10) and B (8, 10), which is the difference of the x-coordinates. length = $8 - 5 = 3$ units

The length between C $(-2, 3)$ and D $(d, 3)$ is:

length = $d - (-2) = d + 2$ units

AB=CD (Property of Parallelogram: Opposite sides are congruent)

$d + 2 = 3 \rightarrow d = 3 - 2 = 1$

18) Answer: A

if $\frac{5}{7}$ cup of nuts =1 cake, to determine how many cakes are made with $4\frac{2}{7}$ cups of nuts, we must first divide $4\frac{2}{7}$ by $\frac{5}{7}$:

$4\frac{2}{7} \div \frac{5}{7} = \frac{30}{7} \div \frac{5}{7} = \frac{30}{7} \times \frac{7}{5} = \frac{210}{35} = 6$

19) Answer: B

Decide by which conversion factor you need to multiply the above fraction in order to eliminate the "m" from the numerator. The conversion equation relating meters, and millimeters is given in this problem. $\left(\frac{100\ cm}{1\ m}\right)$, and $\left(\frac{10\ mm}{1\ cm}\right)$

$\left(\frac{19\ m}{1}\right)\left(\frac{100\ cm}{1\ m}\right)\left(\frac{10\ mm}{1\ cm}\right) = 19{,}000\ mm$

20) Answer: D

$n = 1 \rightarrow 3n - 2 = 3(1) - 2 = 1$

$n = 2 \rightarrow 3n - 2 = 3(2) - 2 = 4$

$n = 3 \rightarrow 3n - 2 = 3(3) - 2 = 7$

WWW.MathNotion.com

$n = 4 \rightarrow 3n - 2 = 3(4) - 2 = 10$

$n = 5 \rightarrow 3n - 2 = 3(5) - 2 = 13$

21) Answer: B

Whole numbers are a subset of integers.

All whole numbers are integers. Some, but not all, integers are whole numbers.

Integers are a subset of rational numbers.

Integer can be positive, negative or zero. All integers are rational numbers.

Some, but not all, rational numbers are integers.

22) Answer: A

Plug in 167 for F and then solve for C.

$C = \frac{5}{9}(F - 32) \Rightarrow C = \frac{5}{9}(167-32) \Rightarrow C = \frac{5}{9}(135) = 75$

23) Answer: A

Let's find the mode, mean (average), and median of the number of minutes for each school.

Number of Minutes for school 1: 50, 60, 70, 70, 70, 80, 100, 110, 110, 120

Mean(average) $= \frac{sum of terms}{number of terms} = \frac{50+ 60+ 70+70+70+80+100+ 110+ 110+,120}{10} = \frac{840}{10} = 84$

Median is the number in the middle. Since there are an even number of items in the resulting list, the median is the average of the two middle numbers.

Median of the data is $(70 + 80) \div 2 = 75$

Mode is the number which appears most often in a set of numbers. Therefore, there is no mode in the set of numbers. Mode is: 70

Number of Minutes for school 2: 50, 60, 80, 80, 90, 100, 110, 110, 110, 120

Mean $= \frac{50+60+80+80+90+100+ 110+ 110+ 110+ 120}{10} = \frac{910}{10} = 91$

Median: $(90 + 100) \div 2 = 95$

Mode: 110

24) Answer: D

$|-57| = 57 > -57$

25) Answer: C

Solve for the sum of 4 numbers.

average = $\frac{\text{sum of terms}}{\text{number of terms}}$ ⇒ $32 = \frac{\text{sum of 4 numbers}}{4}$ ⇒ sum of 4 numbers = $32 \times 4 = 128$

The sum of 4 numbers is 128. If a fifth number 47 is added, then the sum of 5 numbers is: $128 + 47 = 175$

average = $\frac{\text{sum of terms}}{\text{number of terms}} = \frac{175}{5} = 35$

26) Answer: D

The intersecting x- and y-axes divide the coordinate plane into four sections. These four sections are called quadrants.

Point A has a positive x-coordinate and negative y-coordinate, and point B has opposite coordinate of point A. Then point B has the negative x-coordinate and positive y-coordinate. This coordinate related to the fourth quadrant (Q_4).

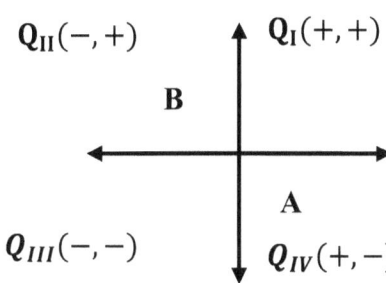

27) Answer: D

The volume of a prism is the number of unit cubes, or cubic units, needed to fill the prism. Then, Volume of a box:

V = length × width × height = $15 \times 9 \times 7 = 945$

28) Answer: B

In mathematics, a percentage is a number or ratio expressed as a fraction of 100.

i.e., $5.6\% = \frac{5.6}{100} = 0.056$

29) Answer: B

$\frac{6}{7} = \frac{36}{y} \rightarrow 6y = 36 \times 7 \rightarrow y = \frac{252}{6} = 42$

30) Answer: A

$\frac{7}{25} \times 100 = 28\%$

SBAC Practice Test 2
Answers and Explanations

1) Answer: B

A point located on one of the axes is not considered to be in a quadrant. It is simply on one of the axes. Whenever the x-coordinate is 0, the point is located on the y-axis. Similarly, any point that has a y-coordinate of 0 will be located on the x-axis

2) Answer: D

Convert all measurement to km, then add.

$3,260\ m = 3.26 km$; and $8,000\ cm = 80\ m = 0.08\ km$

$0.25\ km\ +\ 3,260\ m\ +\ 8,000\ cm\ = 0.25 + 3.26 + 0.08 = 3.59$

3) Answer: D

The base is the value of the number being multiplied, in this case 5 and 2

and the exponent is the number of times, the number is being multiplied. In this case 4 and 2.

4) Answer: A

An isosceles trapezoid has two sides that are the same length and those are not the bases, so the bases are 14 and 20.

The area of the trapezoid then is: $A = \left(\frac{b_1+b_2}{2}\right)h = \left(\frac{14+20}{2}\right)16 = 272$

5) Answer: A

To locate a non-whole number on a number line, we can divide the distance between two whole numbers into fractional parts and then count the number of parts. The segment between 0 and -1 can be partitioned into 20 equal parts. From 0, we can count 5 of the twentieths to locate $-\frac{1}{4} = -\frac{5}{20}$ on the number line.

6) Answer: D

The pattern is: $23, 16, 9, 2, -5, -12$

SBAC Math Practice Grade 6

7) Answer: A

The median of a set of data is the value located in the middle of the data set. To find median, first list numbers in order from smallest to largest:

14, 14, 15, 17, 21, 22

Since there are an even number of items in the resulting list, the median is the average of the two middle numbers.

Median= $(15 + 17) \div 2 = 16$

8) Answer: D

Write this problem as a rate: $7 : 5 = 33 : t \rightarrow \frac{7}{33} = \frac{5}{t}$.

9) Answer: B

$(x + 50°) + 35° + 61° = 180° \rightarrow x + 146° = 180°$

$\rightarrow x = 180° - 146° = 34° \Rightarrow x = 34°$

$\angle B = (x + 50°) = (34° + 50°) = 84°$

10) Answer: C

The perimeter is the total distance around the outside, which can be found by adding together the length of each side. Or as a formula:

Perimeter: $= a + b + c \rightarrow P = c + d + d = 2d + c$

11) Answer: D

Compare each option for $p = 2.3$:

A. $4p < 11.5 \rightarrow 4(2.3) < 11.5 \rightarrow 9.2 \not< 11.5$

B. $15.9 \leq 5.2p \rightarrow 15.9 \leq 5.2(2.3) \rightarrow 15.9 \not\leq 11.96$

C. $7p > 18.6 \rightarrow 7(2.3) > 18.6 \rightarrow 16.1 \not> 18.6$

D. $7.82 \geq 3.4p \rightarrow 7.82 \geq 3.4(2.3) \rightarrow 7.82 \geq 7.82$

12) Answer: D

Let's check the options provided:

A. $\frac{3}{8} \div \frac{3}{5} = \frac{3}{8} \times \frac{5}{3} = \frac{15}{24} = \frac{5}{8} \not< \frac{3}{8}$

B. $\frac{3}{8} \div \frac{3}{4} = \frac{3}{8} \times \frac{4}{3} = \frac{12}{24} = \frac{1}{2} \rightarrow \frac{1}{2} \not< \frac{3}{8}$

C. $\frac{3}{8} \div \frac{1}{4} = \frac{3}{8} \times \frac{4}{1} = \frac{12}{8} = \frac{3}{2} \rightarrow \frac{3}{2} \not< \frac{3}{8}$

D. $\frac{3}{8} \div 3 = \frac{3}{8} \times \frac{1}{3} = \frac{3}{24} = \frac{1}{8} \rightarrow \frac{2}{7} < \frac{3}{8}$

13) Answer: C

The formula for the area of a parallelogram is base times height. $A = b \cdot h$

$A = 12 \times 6 = 72 \, cm^2$

14) Answer: D

A positive number has a negative number for its opposite. A negative number has a positive number for its opposite. The opposite of 0 is itself.

15) Answer: C

$\frac{0.6}{7.50} = \frac{1}{x} \rightarrow 0.6x = 7.50 \rightarrow x = \frac{7.50}{0.6} = \frac{750}{60} = 12.5$

16) Answer: B

Use distributive property:

$4(2x - 3) = 4(2x) + 4(-3) = 8x - 12$

17) Answer: A

The area of the complex shape is equal to the sum of the areas of the divided non-overlapping regions. In this case a rectangle and a trapezoid. Then find the area of each figure.

Area of rectangle: $A_1 = wl = b.d$

Area of trapezoid: $A_2 = \frac{1}{2}h(b_1 + b_2) = \frac{1}{2}a(c + d)$

$A = A_1 + A_2 = bd + \frac{1}{2}a(c + d)$

18) Answer: D

$\frac{5}{30} = \frac{9}{x} \rightarrow 5x = 9 \times 30 \rightarrow 5x = 270 \rightarrow x = \frac{270}{5} = 54$

19) Answer: A

The ratio of the area to its circumference can be found using the following steps:

$\frac{A}{C} = \frac{\pi r^2}{2\pi r} = \frac{r}{2} = \frac{9}{2} = 4.5$

20) Answer: C

Mia has x cards; Mia: x

Mia gave one fifth of her cards to Sofia and so Sofia has $(\frac{1}{5})x$

If Sofia gave $\frac{1}{6}$ of her cards to Riley, she is left with $\frac{5}{6}$ of the cards given to her by Mia.

Hence Sofia has: $(\frac{5}{6})(\frac{1}{5}x) = \frac{5}{30}x = \frac{1}{6}x$ cards.

21) Answer: C

The intercepts of a line are the points where the line crosses the horizontal and vertical axes. The y-intercept is the point where the line crosses the y-axis. The point where the line crosses the x-axis is called the x-intercept. Notice that the y-intercept occurs where x = 0, and the x-intercept occurs where y = 0.

22) Answer: C

$$\text{Probability} = \frac{\text{number of desired outcomes}}{\text{number of total outcomes}} = \frac{4}{5+7+4+2} = \frac{4}{18} = \frac{2}{9}$$

23) Answer: B

The sign in front of each number represents which direction to go on the x or y-axis starting at the origin (0,0). If the number is positive you go right on the x-axis or up on the y-axis. If the sign is negative, then you move to the left on the x-axis or down on the y-axis.

To get to (−3,4), move left on the x-axis by 3, then move up on the y-axis by 4.

24) Answer: D

$\frac{72.6}{1.65} = \frac{7,260}{165} = 44$.

25) Answer: A

Write the ratio and solve for CD:

$\frac{MN}{NP} = \frac{CD}{BD} \rightarrow \frac{10}{8} = \frac{x}{2.4} \rightarrow 8x = 10 \times 2.4 \rightarrow 8x = 24 \rightarrow x = \frac{24}{8} = 3$

26) Answer: C

A histogram is a graphic version of a frequency distribution. The heights of the bars correspond to frequency values. Frequency = $45 + 25 + 10 + 5 + 5 = 90$

27) Answer: A

Round to 0 if the numerator is much smaller than the denominator. Round to $\frac{1}{2}$ if the numerator is about half the denominator. Round to 1 if the numerator is nearly equal to the denominator. Then, $\frac{17}{28} + \frac{19}{35} = 1 + 1 = 2$

28) Answer: C

The line separate second and third quartile indicate the median. The lines outside of the box indicate the outer quartiles (first and fourth).

29) Answer: B

The ratio of 4:5 is the same as 16:20.

30) Answer: B

In the order pair (x, y), each element of the x is paired with exactly one element of the y. So, the relation is a $y = 1.6x$, and substitute $x = 5$,

Then $y = 1.6 \times 5 = 8$. The order pair is $(5, 8)$

"End"